Spark 机器学习(影印版)

Nick Pentreath 著

南京　东南大学出版社

图书在版编目（CIP）数据

Spark 机器学习：英文/（英）彭特里思（Pentreath，N.）著. —影印本. —南京：东南大学出版社，2016.1（2017.1 重印）

书名原文：Machine Learning with Spark

ISBN 978-7-5641-6091-3

Ⅰ. ①S… Ⅱ. ①彭… Ⅲ. ①数据处理软件－机器学习－英文 Ⅳ. ①TP274②TP181

中国版本图书馆 CIP 数据核字（2015）第 256607 号

© 2015 by PACKT Publishing Ltd

Reprint of the English Edition, jointly published by PACKT Publishing Ltd and Southeast University Press, 2016. Authorized reprint of the original English edition, 2015 PACKT Publishing Ltd, the owner of all rights to publish and sell the same.

All rights reserved including the rights of reproduction in whole or in part in any form.

英文原版由 PACKT Publishing Ltd 出版 2015。

英文影印版由东南大学出版社出版 2016。此影印版的出版和销售得到出版权和销售权的所有者——PACKT Publishing Ltd 的许可。

版权所有，未得书面许可，本书的任何部分和全部不得以任何形式重制。

Spark 机器学习（影印版）

出版发行：东南大学出版社
地　　址：南京四牌楼 2 号　　邮编：210096
出 版 人：江建中
网　　址：http://www.seupress.com
电子邮件：press@seupress.com
印　　刷：江苏凤凰数码印务有限公司
开　　本：787 毫米×980 毫米　　16 开本
印　　张：21
字　　数：411 千字
版　　次：2016 年 1 月第 1 版
印　　次：2017 年 1 月第 2 次印刷
书　　号：ISBN 978-7-5641-6091-3
定　　价：68.00 元

本社图书若有印装质量问题，请直接与营销部联系。电话（传真）：025-83791830

Credits

Author
Nick Pentreath

Reviewers
Andrea Mostosi
Hao Ren
Krishna Sankar

Commissioning Editor
Rebecca Youé

Acquisition Editor
Rebecca Youé

Content Development Editor
Susmita Sabat

Technical Editors
Vivek Arora
Pankaj Kadam

Copy Editor
Karuna Narayanan

Project Coordinator
Milton Dsouza

Proofreaders
Simran Bhogal
Maria Gould
Ameesha Green
Paul Hindle

Indexer
Priya Sane

Graphics
Sheetal Aute
Abhinash Sahu

Production Coordinator
Nitesh Thakur

Cover Work
Nitesh Thakur

About the Author

Nick Pentreath has a background in financial markets, machine learning, and software development. He has worked at Goldman Sachs Group, Inc.; as a research scientist at the online ad targeting start-up Cognitive Match Limited, London; and led the Data Science and Analytics team at Mxit, Africa's largest social network.

He is a cofounder of Graphflow, a big data and machine learning company focused on user-centric recommendations and customer intelligence. He is passionate about combining commercial focus with machine learning and cutting-edge technology to build intelligent systems that learn from data to add value to the bottom line.

Nick is a member of the Apache Spark Project Management Committee.

Acknowledgments

Writing this book has been quite a rollercoaster ride over the past year, with many ups and downs, late nights, and working weekends. It has also been extremely rewarding to combine my passion for machine learning with my love of the Apache Spark project, and I hope to bring some of this out in this book.

I would like to thank the Packt Publishing team for all their assistance throughout the writing and editing process: Rebecca, Susmita, Sudhir, Amey, Neil, Vivek, Pankaj, and everyone who worked on the book.

Thanks also go to Debora Donato at StumbleUpon for assistance with data- and legal-related queries.

Writing a book like this can be a somewhat lonely process, so it is incredibly helpful to get the feedback of reviewers to understand whether one is headed in the right direction (and what course adjustments need to be made). I'm deeply grateful to Andrea Mostosi, Hao Ren, and Krishna Sankar for taking the time to provide such detailed and critical feedback.

I could not have gotten through this project without the unwavering support of all my family and friends, especially my wonderful wife, Tammy, who will be glad to have me back in the evenings and on weekends once again. Thank you all!

Finally, thanks to all of you reading this; I hope you find it useful!

About the Reviewers

Andrea Mostosi is a technology enthusiast. An innovation lover since he was a child, he started a professional job in 2003 and worked on several projects, playing almost every role in the computer science environment. He is currently the CTO at The Fool, a company that tries to make sense of web and social data. During his free time, he likes traveling, running, cooking, biking, and coding.

> I would like to thank my geek friends: Simone M, Daniele V, Luca T, Luigi P, Michele N, Luca O, Luca B, Diego C, and Fabio B. They are the smartest people I know, and comparing myself with them has always pushed me to be better.

Hao Ren is a software developer who is passionate about Scala, distributed systems, machine learning, and Apache Spark. He was an exchange student at EPFL when he learned about Scala in 2012. He is currently working in Paris as a backend and data engineer for ClaraVista—a company that focuses on high-performance marketing. His work responsibility is to build a Spark-based platform for purchase prediction and a new recommender system.

Besides programming, he enjoys running, swimming, and playing basketball and badminton. You can learn more at his blog http://www.invkrh.me.

Krishna Sankar is a chief data scientist at BlackArrow, where he is focusing on enhancing user experience via inference, intelligence, and interfaces. Earlier stints include working as a principal architect and data scientist at Tata America International Corporation, director of data science at a bioinformatics start-up company, and as a distinguished engineer at Cisco Systems, Inc. He has spoken at various conferences about data science (http://goo.gl/9pyJMH), machine learning (http://goo.gl/sSem2Y), and social media analysis (http://goo.gl/D9YpVQ). He has also been a guest lecturer at the Naval Postgraduate School. He has written a few books on Java, wireless LAN security, Web 2.0, and now on Spark. His other passion is LEGO robotics. Earlier in April, he was at the St. Louis FLL World Competition as a robots design judge.

www.PacktPub.com

Support files, eBooks, discount offers, and more

For support files and downloads related to your book, please visit www.PacktPub.com.

Did you know that Packt offers eBook versions of every book published, with PDF and ePub files available? You can upgrade to the eBook version at www.PacktPub.com and as a print book customer, you are entitled to a discount on the eBook copy. Get in touch with us at service@packtpub.com for more details.

At www.PacktPub.com, you can also read a collection of free technical articles, sign up for a range of free newsletters and receive exclusive discounts and offers on Packt books and eBooks.

https://www2.packtpub.com/books/subscription/packtlib

Do you need instant solutions to your IT questions? PacktLib is Packt's online digital book library. Here, you can search, access, and read Packt's entire library of books.

Why subscribe?

- Fully searchable across every book published by Packt
- Copy and paste, print, and bookmark content
- On demand and accessible via a web browser

Free access for Packt account holders

If you have an account with Packt at www.PacktPub.com, you can use this to access PacktLib today and view 9 entirely free books. Simply use your login credentials for immediate access.

Table of Contents

Preface	**1**
Chapter 1: Getting Up and Running with Spark	**7**
Installing and setting up Spark locally	8
Spark clusters	10
The Spark programming model	11
SparkContext and SparkConf	11
The Spark shell	12
Resilient Distributed Datasets	14
Creating RDDs	15
Spark operations	15
Caching RDDs	18
Broadcast variables and accumulators	19
The first step to a Spark program in Scala	21
The first step to a Spark program in Java	24
The first step to a Spark program in Python	28
Getting Spark running on Amazon EC2	30
Launching an EC2 Spark cluster	31
Summary	35
Chapter 2: Designing a Machine Learning System	**37**
Introducing MovieStream	38
Business use cases for a machine learning system	39
Personalization	40
Targeted marketing and customer segmentation	40
Predictive modeling and analytics	41
Types of machine learning models	41
The components of a data-driven machine learning system	42
Data ingestion and storage	42
Data cleansing and transformation	43

Model training and testing loop	45
Model deployment and integration	45
Model monitoring and feedback	45
Batch versus real time	47
An architecture for a machine learning system	**48**
Practical exercise	49
Summary	**50**
Chapter 3: Obtaining, Processing, and Preparing Data with Spark	**51**
Accessing publicly available datasets	**52**
The MovieLens 100k dataset	54
Exploring and visualizing your data	**55**
Exploring the user dataset	57
Exploring the movie dataset	62
Exploring the rating dataset	64
Processing and transforming your data	**68**
Filling in bad or missing data	69
Extracting useful features from your data	**70**
Numerical features	71
Categorical features	71
Derived features	73
Transforming timestamps into categorical features	73
Text features	75
Simple text feature extraction	76
Normalizing features	80
Using MLlib for feature normalization	81
Using packages for feature extraction	82
Summary	**82**
Chapter 4: Building a Recommendation Engine with Spark	**83**
Types of recommendation models	**84**
Content-based filtering	85
Collaborative filtering	85
Matrix factorization	86
Extracting the right features from your data	**92**
Extracting features from the MovieLens 100k dataset	92
Training the recommendation model	**96**
Training a model on the MovieLens 100k dataset	96
Training a model using implicit feedback data	98
Using the recommendation model	**99**
User recommendations	99
Generating movie recommendations from the MovieLens 100k dataset	99

Item recommendations	102
Generating similar movies for the MovieLens 100k dataset	103
Evaluating the performance of recommendation models	**106**
Mean Squared Error	107
Mean average precision at K	109
Using MLlib's built-in evaluation functions	113
RMSE and MSE	113
MAP	113
Summary	**115**
Chapter 5: Building a Classification Model with Spark	**117**
Types of classification models	**120**
Linear models	120
Logistic regression	122
Linear support vector machines	123
The naïve Bayes model	124
Decision trees	126
Extracting the right features from your data	**128**
Extracting features from the Kaggle/StumbleUpon evergreen classification dataset	128
Training classification models	**130**
Training a classification model on the Kaggle/StumbleUpon evergreen classification dataset	131
Using classification models	**133**
Generating predictions for the Kaggle/StumbleUpon evergreen classification dataset	133
Evaluating the performance of classification models	**134**
Accuracy and prediction error	134
Precision and recall	136
ROC curve and AUC	138
Improving model performance and tuning parameters	**140**
Feature standardization	141
Additional features	144
Using the correct form of data	147
Tuning model parameters	148
Linear models	149
Decision trees	154
The naïve Bayes model	155
Cross-validation	156
Summary	**159**

Chapter 6: Building a Regression Model with Spark — 161
Types of regression models — 162
- Least squares regression — 162
- Decision trees for regression — 163

Extracting the right features from your data — 164
- Extracting features from the bike sharing dataset — 164
  - Creating feature vectors for the linear model — 168
  - Creating feature vectors for the decision tree — 169

Training and using regression models — 170
- Training a regression model on the bike sharing dataset — 171

Evaluating the performance of regression models — 173
- Mean Squared Error and Root Mean Squared Error — 173
- Mean Absolute Error — 174
- Root Mean Squared Log Error — 174
- The R-squared coefficient — 175
- Computing performance metrics on the bike sharing dataset — 175
  - Linear model — 175
  - Decision tree — 176

Improving model performance and tuning parameters — 177
- Transforming the target variable — 177
  - Impact of training on log-transformed targets — 180
- Tuning model parameters — 183
  - Creating training and testing sets to evaluate parameters — 183
  - The impact of parameter settings for linear models — 184
  - The impact of parameter settings for the decision tree — 192

Summary — 195

Chapter 7: Building a Clustering Model with Spark — 197
Types of clustering models — 198
- K-means clustering — 198
  - Initialization methods — 202
  - Variants — 203
- Mixture models — 203
- Hierarchical clustering — 203

Extracting the right features from your data — 204
- Extracting features from the MovieLens dataset — 204
  - Extracting movie genre labels — 205
  - Training the recommendation model — 207
  - Normalization — 207

Training a clustering model — 208
- Training a clustering model on the MovieLens dataset — 208

Making predictions using a clustering model — 210
- Interpreting cluster predictions on the MovieLens dataset — 211
  - Interpreting the movie clusters — 212

Evaluating the performance of clustering models	**216**
Internal evaluation metrics	216
External evaluation metrics	216
Computing performance metrics on the MovieLens dataset	217
Tuning parameters for clustering models	**217**
Selecting K through cross-validation	217
Summary	**219**

Chapter 8: Dimensionality Reduction with Spark — 221

Types of dimensionality reduction	**222**
Principal Components Analysis	222
Singular Value Decomposition	223
Relationship with matrix factorization	224
Clustering as dimensionality reduction	224
Extracting the right features from your data	**225**
Extracting features from the LFW dataset	225
Exploring the face data	226
Visualizing the face data	228
Extracting facial images as vectors	229
Normalization	233
Training a dimensionality reduction model	**234**
Running PCA on the LFW dataset	235
Visualizing the Eigenfaces	236
Interpreting the Eigenfaces	238
Using a dimensionality reduction model	**238**
Projecting data using PCA on the LFW dataset	239
The relationship between PCA and SVD	240
Evaluating dimensionality reduction models	**242**
Evaluating k for SVD on the LFW dataset	242
Summary	**245**

Chapter 9: Advanced Text Processing with Spark — 247

What's so special about text data?	**247**
Extracting the right features from your data	**248**
Term weighting schemes	248
Feature hashing	249
Extracting the TF-IDF features from the 20 Newsgroups dataset	251
Exploring the 20 Newsgroups data	253
Applying basic tokenization	255
Improving our tokenization	256
Removing stop words	258
Excluding terms based on frequency	261
A note about stemming	264
Training a TF-IDF model	264
Analyzing the TF-IDF weightings	266

Table of Contents

Using a TF-IDF model	**268**
Document similarity with the 20 Newsgroups dataset and TF-IDF features	268
Training a text classifier on the 20 Newsgroups dataset using TF-IDF	271
Evaluating the impact of text processing	**273**
Comparing raw features with processed TF-IDF features on the 20 Newsgroups dataset	273
Word2Vec models	**274**
Word2Vec on the 20 Newsgroups dataset	275
Summary	**278**
Chapter 10: Real-time Machine Learning with Spark Streaming	**279**
Online learning	**279**
Stream processing	**281**
An introduction to Spark Streaming	281
Input sources	282
Transformations	282
Actions	284
Window operators	284
Caching and fault tolerance with Spark Streaming	285
Creating a Spark Streaming application	**286**
The producer application	287
Creating a basic streaming application	290
Streaming analytics	293
Stateful streaming	296
Online learning with Spark Streaming	**298**
Streaming regression	298
A simple streaming regression program	299
Creating a streaming data producer	299
Creating a streaming regression model	302
Streaming K-means	305
Online model evaluation	**306**
Comparing model performance with Spark Streaming	306
Summary	**310**
Index	**311**

Preface

In recent years, the volume of data being collected, stored, and analyzed has exploded, in particular in relation to the activity on the Web and mobile devices, as well as data from the physical world collected via sensor networks. While previously large-scale data storage, processing, analysis, and modeling was the domain of the largest institutions such as Google, Yahoo!, Facebook, and Twitter, increasingly, many organizations are being faced with the challenge of how to handle a massive amount of data.

When faced with this quantity of data and the common requirement to utilize it in real time, human-powered systems quickly become infeasible. This has led to a rise in the so-called big data and machine learning systems that learn from this data to make automated decisions.

In answer to the challenge of dealing with ever larger-scale data without any prohibitive cost, new open source technologies emerged at companies such as Google, Yahoo!, Amazon, and Facebook, which aimed at making it easier to handle massive data volumes by distributing data storage and computation across a cluster of computers.

The most widespread of these is Apache Hadoop, which made it significantly easier and cheaper to both store large amounts of data (via the Hadoop Distributed File System, or HDFS) and run computations on this data (via Hadoop MapReduce, a framework to perform computation tasks in parallel across many nodes in a computer cluster).

Preface

However, MapReduce has some important shortcomings, including high overheads to launch each job and reliance on storing intermediate data and results of the computation to disk, both of which make Hadoop relatively ill-suited for use cases of an iterative or low-latency nature. Apache Spark is a new framework for distributed computing that is designed from the ground up to be optimized for low-latency tasks and to store intermediate data and results in memory, thus addressing some of the major drawbacks of the Hadoop framework. Spark provides a clean, functional, and easy-to-understand API to write applications and is fully compatible with the Hadoop ecosystem.

Furthermore, Spark provides native APIs in Scala, Java, and Python. The Scala and Python APIs allow all the benefits of the Scala or Python language, respectively, to be used directly in Spark applications, including using the relevant interpreter for real-time, interactive exploration. Spark itself now provides a toolkit (called MLlib) of distributed machine learning and data mining models that is under heavy development and already contains high-quality, scalable, and efficient algorithms for many common machine learning tasks, some of which we will delve into in this book.

Applying machine learning techniques to massive datasets is challenging, primarily because most well-known machine learning algorithms are not designed for parallel architectures. In many cases, designing such algorithms is not an easy task. The nature of machine learning models is generally iterative, hence the strong appeal of Spark for this use case. While there are many competing frameworks for parallel computing, Spark is one of the few that combines speed, scalability, in-memory processing, and fault tolerance with ease of programming and a flexible, expressive, and powerful API design.

Throughout this book, we will focus on real-world applications of machine learning technology. While we may briefly delve into some theoretical aspects of machine learning algorithms, the book will generally take a practical, applied approach with a focus on using examples and code to illustrate how to effectively use the features of Spark and MLlib, as well as other well-known and freely available packages for machine learning and data analysis, to create a useful machine learning system.

What this book covers

Chapter 1, *Getting Up and Running with Spark*, shows how to install and set up a local development environment for the Spark framework as well as how to create a Spark cluster in the cloud using Amazon EC2. The Spark programming model and API will be introduced, and a simple Spark application will be created using each of Scala, Java, and Python.

Chapter 2, Designing a Machine Learning System, presents an example of a real-world use case for a machine learning system. We will design a high-level architecture for an intelligent system in Spark based on this illustrative use case.

Chapter 3, Obtaining, Processing, and Preparing Data with Spark, details how to go about obtaining data for use in a machine learning system, in particular from various freely and publicly available sources. We will learn how to process, clean, and transform the raw data into features that may be used in machine learning models, using available tools, libraries, and Spark's functionality.

Chapter 4, Building a Recommendation Engine with Spark, deals with creating a recommendation model based on the collaborative filtering approach. This model will be used to recommend items to a given user as well as create lists of items that are similar to a given item. Standard metrics to evaluate the performance of a recommendation model will be covered here.

Chapter 5, Building a Classification Model with Spark, details how to create a model for binary classification as well as how to utilize standard performance-evaluation metrics for classification tasks.

Chapter 6, Building a Regression Model with Spark, shows how to create a model for regression, extending the classification model created in *Chapter 5, Building a Classification Model with Spark*. Evaluation metrics for the performance of regression models will be detailed here.

Chapter 7, Building a Clustering Model with Spark, explores how to create a clustering model as well as how to use related evaluation methodologies. You will learn how to analyze and visualize the clusters generated.

Chapter 8, Dimensionality Reduction with Spark, takes us through methods to extract the underlying structure from and reduce the dimensionality of our data. You will learn some common dimensionality-reduction techniques and how to apply and analyze them, as well as how to use the resulting data representation as input to another machine learning model.

Chapter 9, Advanced Text Processing with Spark, introduces approaches to deal with large-scale text data, including techniques for feature extraction from text and dealing with the very high-dimensional features typical in text data.

Chapter 10, Real-time Machine Learning with Spark Streaming, provides an overview of Spark Streaming and how it fits in with the online and incremental learning approaches to apply machine learning on data streams.

What you need for this book

Throughout this book, we assume that you have some basic experience with programming in Scala, Java, or Python and have some basic knowledge of machine learning, statistics, and data analysis.

Who this book is for

This book is aimed at entry-level to intermediate data scientists, data analysts, software engineers, and practitioners involved in machine learning or data mining with an interest in large-scale machine learning approaches, but who are not necessarily familiar with Spark. You may have some experience of statistics or machine learning software (perhaps including MATLAB, scikit-learn, Mahout, R, Weka, and so on) or distributed systems (perhaps including some exposure to Hadoop).

Conventions

In this book, you will find a number of styles of text that distinguish between different kinds of information. Here are some examples of these styles, and an explanation of their meaning.

Code words in text, database table names, folder names, filenames, file extensions, pathnames, dummy URLs, user input, and Twitter handles are shown as follows: "Spark places user scripts to run Spark in the bin directory."

A block of code is set as follows:

```
val conf = new SparkConf()
.setAppName("Test Spark App")
.setMaster("local[4]")
val sc = new SparkContext(conf)
```

Any command-line input or output is written as follows:

```
>tar xfvz spark-1.2.0-bin-hadoop2.4.tgz
>cd spark-1.2.0-bin-hadoop2.4
```

New terms and **important words** are shown in bold. Words that you see on the screen, in menus or dialog boxes for example, appear in the text like this: "These can be obtained from the AWS homepage by clicking **Account** | **Security Credentials** | **Access Credentials**."

> Warnings or important notes appear in a box like this.

> Tips and tricks appear like this.

Reader feedback

Feedback from our readers is always welcome. Let us know what you think about this book—what you liked or may have disliked. Reader feedback is important for us to develop titles that you really get the most out of.

To send us general feedback, simply send an e-mail to feedback@packtpub.com, and mention the book title through the subject of your message.

If there is a topic that you have expertise in and you are interested in either writing or contributing to a book, see our author guide on www.packtpub.com/authors.

Customer support

Now that you are the proud owner of a Packt book, we have a number of things to help you to get the most from your purchase.

Downloading the example code

You can download the example code files for all Packt books you have purchased from your account at http://www.packtpub.com. If you purchased this book elsewhere, you can visit http://www.packtpub.com/support and register to have the files e-mailed directly to you.

Errata

Although we have taken every care to ensure the accuracy of our content, mistakes do happen. If you find a mistake in one of our books—maybe a mistake in the text or the code—we would be grateful if you would report this to us. By doing so, you can save other readers from frustration and help us improve subsequent versions of this book. If you find any errata, please report them by visiting `http://www.packtpub.com/support`, selecting your book, clicking on the **Errata Submission Form** link, and entering the details of your errata. Once your errata are verified, your submission will be accepted and the errata will be uploaded to our website or added to any list of existing errata under the Errata section of that title.

To view the previously submitted errata, go to `https://www.packtpub.com/books/content/support` and enter the name of the book in the search field. The required information will appear under the **Errata** section.

Piracy

Piracy of copyright material on the Internet is an ongoing problem across all media. At Packt, we take the protection of our copyright and licenses very seriously. If you come across any illegal copies of our works, in any form, on the Internet, please provide us with the location address or website name immediately so that we can pursue a remedy.

Please contact us at `copyright@packtpub.com` with a link to the suspected pirated material.

We appreciate your help in protecting our authors, and our ability to bring you valuable content.

Questions

You can contact us at `questions@packtpub.com` if you are having a problem with any aspect of the book, and we will do our best to address it.

Getting Up and Running with Spark

Apache Spark is a framework for distributed computing; this framework aims to make it simpler to write programs that run in parallel across many nodes in a cluster of computers. It tries to abstract the tasks of resource scheduling, job submission, execution, tracking, and communication between nodes, as well as the low-level operations that are inherent in parallel data processing. It also provides a higher level API to work with distributed data. In this way, it is similar to other distributed processing frameworks such as Apache Hadoop; however, the underlying architecture is somewhat different.

Spark began as a research project at the University of California, Berkeley. The university was focused on the use case of distributed machine learning algorithms. Hence, it is designed from the ground up for high performance in applications of an iterative nature, where the same data is accessed multiple times. This performance is achieved primarily through caching datasets in memory, combined with low latency and overhead to launch parallel computation tasks. Together with other features such as fault tolerance, flexible distributed-memory data structures, and a powerful functional API, Spark has proved to be broadly useful for a wide range of large-scale data processing tasks, over and above machine learning and iterative analytics.

> For more background on Spark, including the research papers underlying Spark's development, see the project's history page at http://spark.apache.org/community.html#history.

Spark runs in four modes:

- The standalone local mode, where all Spark processes are run within the same **Java Virtual Machine (JVM)** process
- The standalone cluster mode, using Spark's own built-in job-scheduling framework
- Using Mesos, a popular open source cluster-computing framework
- Using YARN (commonly referred to as NextGen MapReduce), a Hadoop-related cluster-computing and resource-scheduling framework

In this chapter, we will:

- Download the Spark binaries and set up a development environment that runs in Spark's standalone local mode. This environment will be used throughout the rest of the book to run the example code.
- Explore Spark's programming model and API using Spark's interactive console.
- Write our first Spark program in Scala, Java, and Python.
- Set up a Spark cluster using Amazon's **Elastic Cloud Compute (EC2)** platform, which can be used for large-sized data and heavier computational requirements, rather than running in the local mode.

> Spark can also be run on Amazon's Elastic MapReduce service using custom bootstrap action scripts, but this is beyond the scope of this book. The following article is a good reference guide: http://aws.amazon.com/articles/Elastic-MapReduce/4926593393724923.
>
> At the time of writing this book, the article covers running Spark Version 1.1.0.

If you have previous experience in setting up Spark and are familiar with the basics of writing a Spark program, feel free to skip this chapter.

Installing and setting up Spark locally

Spark can be run using the built-in standalone cluster scheduler in the local mode. This means that all the Spark processes are run within the same JVM—effectively, a single, multithreaded instance of Spark. The local mode is very useful for prototyping, development, debugging, and testing. However, this mode can also be useful in real-world scenarios to perform parallel computation across multiple cores on a single computer.

As Spark's local mode is fully compatible with the cluster mode, programs written and tested locally can be run on a cluster with just a few additional steps.

The first step in setting up Spark locally is to download the latest version (at the time of writing this book, the version is 1.2.0). The download page of the Spark project website, found at http://spark.apache.org/downloads.html, contains links to download various versions as well as to obtain the latest source code via GitHub.

> The Spark project documentation website at http://spark.apache.org/docs/latest/ is a comprehensive resource to learn more about Spark. We highly recommend that you explore it!

Spark needs to be built against a specific version of Hadoop in order to access **Hadoop Distributed File System** (**HDFS**) as well as standard and custom Hadoop input sources. The download page provides prebuilt binary packages for Hadoop 1, CDH4 (Cloudera's Hadoop Distribution), MapR's Hadoop distribution, and Hadoop 2 (YARN). Unless you wish to build Spark against a specific Hadoop version, we recommend that you download the prebuilt Hadoop 2.4 package from an Apache mirror using this link: http://www.apache.org/dyn/closer.cgi/spark/spark-1.2.0/spark-1.2.0-bin-hadoop2.4.tgz.

Spark requires the Scala programming language (version 2.10.4 at the time of writing this book) in order to run. Fortunately, the prebuilt binary package comes with the Scala runtime packages included, so you don't need to install Scala separately in order to get started. However, you will need to have a **Java Runtime Environment** (**JRE**) or **Java Development Kit** (**JDK**) installed (see the software and hardware list in this book's code bundle for installation instructions).

Once you have downloaded the Spark binary package, unpack the contents of the package and change into the newly created directory by running the following commands:

```
>tar xfvz spark-1.2.0-bin-hadoop2.4.tgz
>cd spark-1.2.0-bin-hadoop2.4
```

Spark places user scripts to run Spark in the `bin` directory. You can test whether everything is working correctly by running one of the example programs included in Spark:

```
>./bin/run-example org.apache.spark.examples.SparkPi
```

Getting Up and Running with Spark

This will run the example in Spark's local standalone mode. In this mode, all the Spark processes are run within the same JVM, and Spark uses multiple threads for parallel processing. By default, the preceding example uses a number of threads equal to the number of cores available on your system. Once the program is finished running, you should see something similar to the following lines near the end of the output:

```
...
14/11/27 20:58:47 INFO SparkContext: Job finished: reduce at SparkPi.
scala:35, took 0.723269 s
Pi is roughly 3.1465
...
```

To configure the level of parallelism in the local mode, you can pass in a `master` parameter of the `local[N]` form, where N is the number of threads to use. For example, to use only two threads, run the following command instead:

```
>MASTER=local[2] ./bin/run-example org.apache.spark.examples.SparkPi
```

Spark clusters

A Spark cluster is made up of two types of processes: a driver program and multiple executors. In the local mode, all these processes are run within the same JVM. In a cluster, these processes are usually run on separate nodes.

For example, a typical cluster that runs in Spark's standalone mode (that is, using Spark's built-in cluster-management modules) will have:

- A master node that runs the Spark standalone master process as well as the driver program
- A number of worker nodes, each running an executor process

While we will be using Spark's local standalone mode throughout this book to illustrate concepts and examples, the same Spark code that we write can be run on a Spark cluster. In the preceding example, if we run the code on a Spark standalone cluster, we could simply pass in the URL for the master node as follows:

```
>MASTER=spark://IP:PORT ./bin/run-example org.apache.spark.examples.
SparkPi
```

Here, IP is the IP address, and PORT is the port of the Spark master. This tells Spark to run the program on the cluster where the Spark master process is running.

A full treatment of Spark's cluster management and deployment is beyond the scope of this book. However, we will briefly teach you how to set up and use an Amazon EC2 cluster later in this chapter.

> For an overview of the Spark cluster-application deployment, take a look at the following links:
> - `http://spark.apache.org/docs/latest/cluster-overview.html`
> - `http://spark.apache.org/docs/latest/submitting-applications.html`

The Spark programming model

Before we delve into a high-level overview of Spark's design, we will introduce the `SparkContext` object as well as the Spark shell, which we will use to interactively explore the basics of the Spark programming model.

> While this section provides a brief overview and examples of using Spark, we recommend that you read the following documentation to get a detailed understanding:
> - Spark Quick Start: `http://spark.apache.org/docs/latest/quick-start.html`
> - *Spark Programming guide*, which covers Scala, Java, and Python: `http://spark.apache.org/docs/latest/programming-guide.html`

SparkContext and SparkConf

The starting point of writing any Spark program is `SparkContext` (or `JavaSparkContext` in Java). `SparkContext` is initialized with an instance of a `SparkConf` object, which contains various Spark cluster-configuration settings (for example, the URL of the master node).

Once initialized, we will use the various methods found in the `SparkContext` object to create and manipulate distributed datasets and shared variables. The Spark shell (in both Scala and Python, which is unfortunately not supported in Java) takes care of this context initialization for us, but the following lines of code show an example of creating a context running in the local mode in Scala:

```
val conf = new SparkConf()
.setAppName("Test Spark App")
.setMaster("local[4]")
val sc = new SparkContext(conf)
```

This creates a context running in the local mode with four threads, with the name of the application set to `Test Spark App`. If we wish to use default configuration values, we could also call the following simple constructor for our `SparkContext` object, which works in exactly the same way:

```
val sc = new SparkContext("local[4]", "Test Spark App")
```

> **Downloading the example code**
>
> You can download the example code files for all Packt books you have purchased from your account at http://www.packtpub.com. If you purchased this book elsewhere, you can visit http://www.packtpub.com/support and register to have the files e-mailed directly to you.

The Spark shell

Spark supports writing programs interactively using either the Scala or Python REPL (that is, the **Read-Eval-Print-Loop**, or interactive shell). The shell provides instant feedback as we enter code, as this code is immediately evaluated. In the Scala shell, the return result and type is also displayed after a piece of code is run.

To use the Spark shell with Scala, simply run `./bin/spark-shell` from the Spark base directory. This will launch the Scala shell and initialize `SparkContext`, which is available to us as the Scala value, `sc`. Your console output should look similar to the following screenshot:

```
Nicks-MacBook-Pro:spark-1.2.0-bin-hadoop2.4 Nick$ ./bin/spark-shell
Using Spark's default log4j profile: org/apache/spark/log4j-defaults.properties
14/11/27 22:02:26 INFO SecurityManager: Changing view acls to: Nick
14/11/27 22:02:26 INFO SecurityManager: Changing modify acls to: Nick
14/11/27 22:02:26 INFO SecurityManager: SecurityManager: authentication disabled; ui acls disabled; users with view per
missions: Set(Nick); users with modify permissions: Set(Nick)
14/11/27 22:02:26 INFO HttpServer: Starting HTTP Server
14/11/27 22:02:26 INFO Utils: Successfully started service 'HTTP class server' on port 55288.
Welcome to
      ____              __
     / __/__  ___ _____/ /__
    _\ \/ _ \/ _ `/ __/  '_/
   /___/ .__/\_,_/_/ /_/\_\   version 1.2.0
      /_/

Using Scala version 2.10.4 (Java HotSpot(TM) 64-Bit Server VM, Java 1.7.0_60)
Type in expressions to have them evaluated.
Type :help for more information.
14/11/27 22:02:30 WARN Utils: Your hostname, Nicks-MacBook-Pro.local resolves to a loopback address: 127.0.0.1; using 1
0.0.0.7 instead (on interface en0)
14/11/27 22:02:30 WARN Utils: Set SPARK_LOCAL_IP if you need to bind to another address
14/11/27 22:02:30 INFO SecurityManager: Changing view acls to: Nick
14/11/27 22:02:30 INFO SecurityManager: Changing modify acls to: Nick
14/11/27 22:02:30 INFO SecurityManager: SecurityManager: authentication disabled; ui acls disabled; users with view per
missions: Set(Nick); users with modify permissions: Set(Nick)
14/11/27 22:02:31 INFO Slf4jLogger: Slf4jLogger started
14/11/27 22:02:31 INFO Remoting: Starting remoting
14/11/27 22:02:31 INFO Remoting: Remoting started; listening on addresses :[akka.tcp://sparkDriver@10.0.0.7:55290]
14/11/27 22:02:31 INFO Utils: Successfully started service 'sparkDriver' on port 55290.
14/11/27 22:02:31 INFO SparkEnv: Registering MapOutputTracker
14/11/27 22:02:31 INFO SparkEnv: Registering BlockManagerMaster
14/11/27 22:02:31 INFO DiskBlockManager: Created local directory at /var/folders/_l/06wxljt13wqgm7r08jlc44_r0000gn/T/sp
ark-local-20141127220231-634b
14/11/27 22:02:31 INFO MemoryStore: MemoryStore started with capacity 265.4 MB
14/11/27 22:02:31 WARN NativeCodeLoader: Unable to load native-hadoop library for your platform... using builtin-java c
lasses where applicable
14/11/27 22:02:31 INFO HttpFileServer: HTTP File server directory is /var/folders/_l/06wxljt13wqgm7r08jlc44_r0000gn/T/s
park-0595fd59-f23f-4b83-8cda-5b7b68534335
14/11/27 22:02:31 INFO HttpServer: Starting HTTP Server
14/11/27 22:02:31 INFO Utils: Successfully started service 'HTTP file server' on port 55291.
14/11/27 22:02:32 INFO Utils: Successfully started service 'SparkUI' on port 4040.
14/11/27 22:02:32 INFO SparkUI: Started SparkUI at http://10.0.0.7:4040
14/11/27 22:02:32 INFO Executor: Using REPL class URI: http://10.0.0.7:55288
14/11/27 22:02:32 INFO AkkaUtils: Connecting to HeartbeatReceiver: akka.tcp://sparkDriver@10.0.0.7:55290/user/Heartbeat
Receiver
14/11/27 22:02:32 INFO NettyBlockTransferService: Server created on 55292
14/11/27 22:02:32 INFO BlockManagerMaster: Trying to register BlockManager
14/11/27 22:02:32 INFO BlockManagerMasterActor: Registering block manager localhost:55292 with 265.4 MB RAM, BlockManag
erId(<driver>, localhost, 55292)
14/11/27 22:02:32 INFO BlockManagerMaster: Registered BlockManager
14/11/27 22:02:32 INFO SparkILoop: Created spark context..
Spark context available as sc.

scala>
```

To use the Python shell with Spark, simply run the `./bin/pyspark` command. Like the Scala shell, the Python `SparkContext` object should be available as the Python variable `sc`. You should see an output similar to the one shown in this screenshot:

```
Nicks-MacBook-Pro:spark-1.2.0-bin-hadoop2.4 Nick$ ./bin/pyspark
Python 2.7.8 |Anaconda 2.0.1 (x86_64)| (default, Aug 21 2014, 15:21:46)
[GCC 4.2.1 (Apple Inc. build 5577)] on darwin
Type "help", "copyright", "credits" or "license" for more information.
Anaconda is brought to you by Continuum Analytics.
Please check out: http://continuum.io/thanks and https://binstar.org
Using Spark's default log4j profile: org/apache/spark/log4j-defaults.properties
14/11/27 22:05:24 WARN Utils: Your hostname, Nicks-MacBook-Pro.local resolves to a loopback address: 127.0.0.1; using 1
0.0.0.7 instead (on interface en0)
14/11/27 22:05:24 WARN Utils: Set SPARK_LOCAL_IP if you need to bind to another address
14/11/27 22:05:24 INFO SecurityManager: Changing view acls to: Nick
14/11/27 22:05:24 INFO SecurityManager: Changing modify acls to: Nick
14/11/27 22:05:24 INFO SecurityManager: SecurityManager: authentication disabled; ui acls disabled; users with view per
missions: Set(Nick); users with modify permissions: Set(Nick)
14/11/27 22:05:24 INFO Slf4jLogger: Slf4jLogger started
14/11/27 22:05:24 INFO Remoting: Starting remoting
14/11/27 22:05:25 INFO Remoting: Remoting started; listening on addresses :[akka.tcp://sparkDriver@10.0.0.7:55313]
14/11/27 22:05:25 INFO Utils: Successfully started service 'sparkDriver' on port 55313.
14/11/27 22:05:25 INFO SparkEnv: Registering MapOutputTracker
14/11/27 22:05:25 INFO SparkEnv: Registering BlockManagerMaster
14/11/27 22:05:25 INFO DiskBlockManager: Created local directory at /var/folders/_l/06wxljt13wqgm7r08jlc44_r0000gn/T/sp
ark-local-20141127220525-7631
14/11/27 22:05:25 INFO MemoryStore: MemoryStore started with capacity 265.4 MB
14/11/27 22:05:25 WARN NativeCodeLoader: Unable to load native-hadoop library for your platform... using builtin-java c
lasses where applicable
14/11/27 22:05:25 INFO HttpFileServer: HTTP File server directory is /var/folders/_l/06wxljt13wqgm7r08jlc44_r0000gn/T/s
park-e5b50a14-c102-40bd-a04a-ba69485dfbea
14/11/27 22:05:25 INFO HttpServer: Starting HTTP Server
14/11/27 22:05:25 INFO Utils: Successfully started service 'HTTP file server' on port 55314.
14/11/27 22:05:25 INFO Utils: Successfully started service 'SparkUI' on port 4040.
14/11/27 22:05:25 INFO SparkUI: Started SparkUI at http://10.0.0.7:4040
14/11/27 22:05:25 INFO AkkaUtils: Connecting to HeartbeatReceiver: akka.tcp://sparkDriver@10.0.0.7:55313/user/Heartbeat
Receiver
14/11/27 22:05:25 INFO NettyBlockTransferService: Server created on 55315
14/11/27 22:05:25 INFO BlockManagerMaster: Trying to register BlockManager
14/11/27 22:05:25 INFO BlockManagerMasterActor: Registering block manager localhost:55315 with 265.4 MB RAM, BlockManag
erId(<driver>, localhost, 55315)
14/11/27 22:05:25 INFO BlockManagerMaster: Registered BlockManager
Welcome to
      ____              __
     / __/__  ___ _____/ /__
    _\ \/ _ \/ _ `/ __/  '_/
   /__ / .__/\_,_/_/ /_/\_\   version 1.2.0
      /_/

Using Python version 2.7.8 (default, Aug 21 2014 15:21:46)
SparkContext available as sc.
>>>
```

Resilient Distributed Datasets

The core of Spark is a concept called the **Resilient Distributed Dataset** (**RDD**). An RDD is a collection of "records" (strictly speaking, objects of some type) that is distributed or partitioned across many nodes in a cluster (for the purposes of the Spark local mode, the single multithreaded process can be thought of in the same way). An RDD in Spark is fault-tolerant; this means that if a given node or task fails (for some reason other than erroneous user code, such as hardware failure, loss of communication, and so on), the RDD can be reconstructed automatically on the remaining nodes and the job will still complete.

Creating RDDs

RDDs can be created from existing collections, for example, in the Scala Spark shell that you launched earlier:

```
val collection = List("a", "b", "c", "d", "e")
val rddFromCollection = sc.parallelize(collection)
```

RDDs can also be created from Hadoop-based input sources, including the local filesystem, HDFS, and Amazon S3. A Hadoop-based RDD can utilize any input format that implements the Hadoop InputFormat interface, including text files, other standard Hadoop formats, HBase, Cassandra, and many more. The following code is an example of creating an RDD from a text file located on the local filesystem:

```
val rddFromTextFile = sc.textFile("LICENSE")
```

The preceding textFile method returns an RDD where each record is a String object that represents one line of the text file.

Spark operations

Once we have created an RDD, we have a distributed collection of records that we can manipulate. In Spark's programming model, operations are split into transformations and actions. Generally speaking, a transformation operation applies some function to all the records in the dataset, changing the records in some way. An action typically runs some computation or aggregation operation and returns the result to the driver program where SparkContext is running.

Spark operations are functional in style. For programmers familiar with functional programming in Scala or Python, these operations should seem natural. For those without experience in functional programming, don't worry; the Spark API is relatively easy to learn.

One of the most common transformations that you will use in Spark programs is the map operator. This applies a function to each record of an RDD, thus *mapping* the input to some new output. For example, the following code fragment takes the RDD we created from a local text file and applies the size function to each record in the RDD. Remember that we created an RDD of Strings. Using map, we can transform each string to an integer, thus returning an RDD of Ints:

```
val intsFromStringsRDD = rddFromTextFile.map(line => line.size)
```

Getting Up and Running with Spark

You should see output similar to the following line in your shell; this indicates the type of the RDD:

```
intsFromStringsRDD: org.apache.spark.rdd.RDD[Int] = MappedRDD[5] at map at <console>:14
```

In the preceding code, we saw the `=>` syntax used. This is the Scala syntax for an anonymous function, which is a function that is not a named method (that is, one defined using the `def` keyword in Scala or Python, for example).

> While a detailed treatment of anonymous functions is beyond the scope of this book, they are used extensively in Spark code in Scala and Python, as well as in Java 8 (both in examples and real-world applications), so it is useful to cover a few practicalities.
>
> The `line => line.size` syntax means that we are applying a function where the input variable is to the left of the `=>` operator, and the output is the result of the code to the right of the `=>` operator. In this case, the input is `line`, and the output is the result of calling `line.size`. In Scala, this function that maps a string to an integer is expressed as `String => Int`.
>
> This syntax saves us from having to separately define functions every time we use methods such as `map`; this is useful when the function is simple and will only be used once, as in this example.

Now, we can apply a common action operation, `count`, to return the number of records in our RDD:

```
intsFromStringsRDD.count
```

The result should look something like the following console output:

```
14/01/29 23:28:28 INFO SparkContext: Starting job: count at <console>:17
...
14/01/29 23:28:28 INFO SparkContext: Job finished: count at <console>:17, took 0.019227 s
res4: Long = 398
```

Perhaps we want to find the average length of each line in this text file. We can first use the `sum` function to add up all the lengths of all the records and then divide the sum by the number of records:

```
val sumOfRecords = intsFromStringsRDD.sum
val numRecords = intsFromStringsRDD.count
val aveLengthOfRecord = sumOfRecords / numRecords
```

The result will be as follows:

```
aveLengthOfRecord: Double = 52.06030150753769
```

Spark operations, in most cases, return a new RDD, with the exception of most actions, which return the result of a computation (such as `Long` for `count` and `Double` for `sum` in the preceding example). This means that we can naturally chain together operations to make our program flow more concise and expressive. For example, the same result as the one in the preceding line of code can be achieved using the following code:

```
val aveLengthOfRecordChained = rddFromTextFile.map(line => line.size).
sum / rddFromTextFile.count
```

An important point to note is that Spark transformations are lazy. That is, invoking a transformation on an RDD does not immediately trigger a computation. Instead, transformations are chained together and are effectively only computed when an action is called. This allows Spark to be more efficient by only returning results to the driver when necessary so that the majority of operations are performed in parallel on the cluster.

This means that if your Spark program never uses an action operation, it will never trigger an actual computation, and you will not get any results. For example, the following code will simply return a new RDD that represents the chain of transformations:

```
val transformedRDD = rddFromTextFile.map(line => line.size).
filter(size => size > 10).map(size => size * 2)
```

This returns the following result in the console:

```
transformedRDD: org.apache.spark.rdd.RDD[Int] = MappedRDD[8] at map at <console>:14
```

Notice that no actual computation happens and no result is returned. If we now call an action, such as `sum`, on the resulting RDD, the computation will be triggered:

```
val computation = transformedRDD.sum
```

You will now see that a Spark job is run, and it results in the following console output:

```
...
14/11/27 21:48:21 INFO SparkContext: Job finished: sum at <console>:16, took 0.193513 s
computation: Double = 60468.0
```

Getting Up and Running with Spark

> The complete list of transformations and actions possible on RDDs as well as a set of more detailed examples are available in the Spark programming guide (located at http://spark.apache.org/docs/latest/programming-guide.html#rdd-operations), and the API documentation (the Scala API documentation) is located at http://spark.apache.org/docs/latest/api/scala/index.html#org.apache.spark.rdd.RDD).

Caching RDDs

One of the most powerful features of Spark is the ability to cache data in memory across a cluster. This is achieved through use of the `cache` method on an RDD:

```
rddFromTextFile.cache
```

Calling `cache` on an RDD tells Spark that the RDD should be kept in memory. The first time an action is called on the RDD that initiates a computation, the data is read from its source and put into memory. Hence, the first time such an operation is called, the time it takes to run the task is partly dependent on the time it takes to read the data from the input source. However, when the data is accessed the next time (for example, in subsequent queries in analytics or iterations in a machine learning model), the data can be read directly from memory, thus avoiding expensive I/O operations and speeding up the computation, in many cases, by a significant factor.

If we now call the `count` or `sum` function on our cached RDD, we will see that the RDD is loaded into memory:

```
val aveLengthOfRecordChained = rddFromTextFile.map(line => line.size).
sum / rddFromTextFile.count
```

Indeed, in the following output, we see that the dataset was cached in memory on the first call, taking up approximately 62 KB and leaving us with around 270 MB of memory free:

```
...
14/01/30 06:59:27 INFO MemoryStore: ensureFreeSpace(63454) called with
curMem=32960, maxMem=311387750
14/01/30 06:59:27 INFO MemoryStore: Block rdd_2_0 stored as values to
memory (estimated size 62.0 KB, free 296.9 MB)
14/01/30 06:59:27 INFO BlockManagerMasterActor$BlockManagerInfo: Added
rdd_2_0 in memory on 10.0.0.3:55089 (size: 62.0 KB, free: 296.9 MB)
...
```

Now, we will call the same function again:

```
val aveLengthOfRecordChainedFromCached = rddFromTextFile.map(line =>
line.size).sum / rddFromTextFile.count
```

We will see from the console output that the cached data is read directly from memory:

...

`14/01/30 06:59:34 INFO BlockManager: Found block rdd_2_0 locally`

...

 Spark also allows more fine-grained control over caching behavior. You can use the `persist` method to specify what approach Spark uses to cache data. More information on RDD caching can be found here: http://spark.apache.org/docs/latest/programming-guide.html#rdd-persistence.

Broadcast variables and accumulators

Another core feature of Spark is the ability to create two special types of variables: broadcast variables and accumulators.

A **broadcast variable** is a *read-only* variable that is made available from the driver program that runs the `SparkContext` object to the nodes that will execute the computation. This is very useful in applications that need to make the same data available to the worker nodes in an efficient manner, such as machine learning algorithms. Spark makes creating broadcast variables as simple as calling a method on `SparkContext` as follows:

```
val broadcastAList = sc.broadcast(List("a", "b", "c", "d", "e"))
```

The console output shows that the broadcast variable was stored in memory, taking up approximately 488 bytes, and it also shows that we still have 270 MB available to us:

`14/01/30 07:13:32 INFO MemoryStore: ensureFreeSpace(488) called with curMem=96414, maxMem=311387750`

`14/01/30 07:13:32 INFO MemoryStore: Block broadcast_1 stored as values to memory (estimated size 488.0 B, free 296.9 MB)`

`broadCastAList: org.apache.spark.broadcast.Broadcast[List[String]] = Broadcast(1)`

Getting Up and Running with Spark

A broadcast variable can be accessed from nodes other than the driver program that created it (that is, the worker nodes) by calling `value` on the variable:

```
sc.parallelize(List("1", "2", "3")).map(x => broadcastAList.value ++ x).collect
```

This code creates a new RDD with three records from a collection (in this case, a Scala `List`) of `("1", "2", "3")`. In the `map` function, it returns a new collection with the relevant record from our new RDD appended to the `broadcastAList` that is our broadcast variable.

Notice that we used the `collect` method in the preceding code. This is a Spark *action* that returns the entire RDD to the driver as a Scala (or Python or Java) collection.

We will often use `collect` when we wish to apply further processing to our results locally within the driver program.

> Note that `collect` should generally only be used in cases where we really want to return the full result set to the driver and perform further processing. If we try to call `collect` on a very large dataset, we might run out of memory on the driver and crash our program.
>
> It is preferable to perform as much heavy-duty processing on our Spark cluster as possible, preventing the driver from becoming a bottleneck. In many cases, however, collecting results to the driver is necessary, such as during iterations in many machine learning models.

On inspecting the result, we will see that for each of the three records in our new RDD, we now have a record that is our original broadcasted `List`, with the new element appended to it (that is, there is now either `"1"`, `"2"`, or `"3"` at the end):

```
...
14/01/31 10:15:39 INFO SparkContext: Job finished: collect at
<console>:15, took 0.025806 s
res6: Array[List[Any]] = Array(List(a, b, c, d, e, 1), List(a, b, c, d,
e, 2), List(a, b, c, d, e, 3))
```

An **accumulator** is also a variable that is broadcasted to the worker nodes. The key difference between a broadcast variable and an accumulator is that while the broadcast variable is read-only, the accumulator can be added to. There are limitations to this, that is, in particular, the addition must be an associative operation so that the global accumulated value can be correctly computed in parallel and returned to the driver program. Each worker node can only access and add to its own local accumulator value, and only the driver program can access the global value. Accumulators are also accessed within the Spark code using the `value` method.

> For more details on broadcast variables and accumulators, see the *Shared Variables* section of the *Spark Programming Guide*: http://spark.apache.org/docs/latest/programming-guide.html#shared-variables.

The first step to a Spark program in Scala

We will now use the ideas we introduced in the previous section to write a basic Spark program to manipulate a dataset. We will start with Scala and then write the same program in Java and Python. Our program will be based on exploring some data from an online store, about which users have purchased which products. The data is contained in a **comma-separated-value** (**CSV**) file called `UserPurchaseHistory.csv`, and the contents are shown in the following snippet. The first column of the CSV is the username, the second column is the product name, and the final column is the price:

```
John,iPhone Cover,9.99
John,Headphones,5.49
Jack,iPhone Cover,9.99
Jill,Samsung Galaxy Cover,8.95
Bob,iPad Cover,5.49
```

For our Scala program, we need to create two files: our Scala code and our project build configuration file, using the build tool **Scala Build Tool** (**sbt**). For ease of use, we recommend that you download the sample project code called `scala-spark-app` for this chapter. This code also contains the CSV file under the `data` directory. You will need SBT installed on your system in order to run this example program (we use version 0.13.1 at the time of writing this book).

> Setting up SBT is beyond the scope of this book; however, you can find more information at http://www.scala-sbt.org/release/docs/Getting-Started/Setup.html.

Our SBT configuration file, `build.sbt`, looks like this (note that the empty lines between each line of code are required):

```
name := "scala-spark-app"

version := "1.0"
```

```
scalaVersion := "2.10.4"

libraryDependencies += "org.apache.spark" %% "spark-core" % "1.2.0 "
```

The last line adds the dependency on Spark to our project.

Our Scala program is contained in the `ScalaApp.scala` file. We will walk through the program piece by piece. First, we need to import the required Spark classes:

```
import org.apache.spark.SparkContext
import org.apache.spark.SparkContext._

/**
 * A simple Spark app in Scala
 */
object ScalaApp {
```

In our main method, we need to initialize our `SparkContext` object and use this to access our CSV data file with the `textFile` method. We will then map the raw text by splitting the string on the delimiter character (a comma in this case) and extracting the relevant records for username, product, and price:

```
  def main(args: Array[String]) {
    val sc = new SparkContext("local[2]", "First Spark App")
    // we take the raw data in CSV format and convert it into a
    // set of records of the form (user, product, price)
    val data = sc.textFile("data/UserPurchaseHistory.csv")
      .map(line => line.split(","))
      .map(purchaseRecord => (purchaseRecord(0),
    purchaseRecord(1), purchaseRecord(2)))
```

Now that we have an RDD, where each record is made up of (`user`, `product`, `price`), we can compute various interesting metrics for our store, such as the following ones:

- The total number of purchases
- The number of unique users who purchased
- Our total revenue
- Our most popular product

Let's compute the preceding metrics:

```
// let's count the number of purchases
val numPurchases = data.count()
// let's count how many unique users made purchases
val uniqueUsers = data.map{ case (user, product, price) => user
}.distinct().count()
// let's sum up our total revenue
val totalRevenue = data.map{ case (user, product, price) => price.
toDouble }.sum()
// let's find our most popular product
val productsByPopularity = data
  .map{ case (user, product, price) => (product, 1) }
  .reduceByKey(_ + _)
  .collect()
  .sortBy(-_._2)
val mostPopular = productsByPopularity(0)
```

This last piece of code to compute the most popular product is an example of the *Map/Reduce* pattern made popular by Hadoop. First, we mapped our records of (user, product, price) to the records of (product, 1). Then, we performed a reduceByKey operation, where we summed up the 1s for each unique product.

Once we have this transformed RDD, which contains the number of purchases for each product, we will call collect, which returns the results of the computation to the driver program as a local Scala collection. We will then sort these counts locally (note that in practice, if the amount of data is large, we will perform the sorting in parallel, usually with a Spark operation such as sortByKey).

Finally, we will print out the results of our computations to the console:

```
    println("Total purchases: " + numPurchases)
    println("Unique users: " + uniqueUsers)
    println("Total revenue: " + totalRevenue)
    println("Most popular product: %s with %d purchases".
format(mostPopular._1, mostPopular._2))
  }
}
```

We can run this program by running sbt run in the project's base directory or by running the program in your Scala IDE if you are using one. The output should look similar to the following:

```
...
[info] Compiling 1 Scala source to ...
```

```
[info] Running ScalaApp
...
14/01/30 10:54:40 INFO spark.SparkContext: Job finished: collect at
ScalaApp.scala:25, took 0.045181 s
Total purchases: 5
Unique users: 4
Total revenue: 39.91
Most popular product: iPhone Cover with 2 purchases
```

We can see that we have five purchases from four different users with a total revenue of 39.91. Our most popular product is an iPhone cover with 2 purchases.

The first step to a Spark program in Java

The Java API is very similar in principle to the Scala API. However, while Scala can call the Java code quite easily, in some cases, it is not possible to call the Scala code from Java. This is particularly the case when such Scala code makes use of certain Scala features such as implicit conversions, default parameters, and the Scala reflection API.

Spark makes heavy use of these features in general, so it is necessary to have a separate API specifically for Java that includes Java versions of the common classes. Hence, `SparkContext` becomes `JavaSparkContext`, and `RDD` becomes `JavaRDD`.

Java versions prior to version 8 do not support anonymous functions and do not have succinct syntax for functional-style programming, so functions in the Spark Java API must implement a `WrappedFunction` interface with the `call` method signature. While it is significantly more verbose, we will often create one-off anonymous classes to pass to our Spark operations, which implement this interface and the `call` method, to achieve much the same effect as anonymous functions in Scala.

Spark provides support for Java 8's anonymous function (or *lambda*) syntax. Using this syntax makes a Spark program written in Java 8 look very close to the equivalent Scala program.

In Scala, an RDD of key/value pairs provides special operators (such as `reduceByKey` and `saveAsSequenceFile`, for example) that are accessed automatically via implicit conversions. In Java, special types of `JavaRDD` classes are required in order to access similar functions. These include `JavaPairRDD` to work with key/value pairs and `JavaDoubleRDD` to work with numerical records.

> In this section, we covered the standard Java API syntax. For more details and examples related to working RDDs in Java as well as the Java 8 lambda syntax, see the Java sections of the *Spark Programming Guide* found at http://spark.apache.org/docs/latest/programming-guide.html#rdd-operations.

We will see examples of most of these differences in the following Java program, which is included in the example code of this chapter in the directory named `java-spark-app`. The code directory also contains the CSV data file under the `data` subdirectory.

We will build and run this project with the Maven build tool, which we assume you have installed on your system.

> Installing and setting up Maven is beyond the scope of this book. Usually, Maven can easily be installed using the package manager on your Linux system or HomeBrew or MacPorts on Mac OS X.
>
> Detailed installation instructions can be found here: http://maven.apache.org/download.cgi.

The project contains a Java file called `JavaApp.java`, which contains our program code:

```java
import org.apache.spark.api.java.JavaRDD;
import org.apache.spark.api.java.JavaSparkContext;
import org.apache.spark.api.java.function.DoubleFunction;
import org.apache.spark.api.java.function.Function;
import org.apache.spark.api.java.function.Function2;
import org.apache.spark.api.java.function.PairFunction;
import scala.Tuple2;

import java.util.Collections;
import java.util.Comparator;
import java.util.List;

/**
 * A simple Spark app in Java
 */
public class JavaApp {

    public static void main(String[] args) {
```

As in our Scala example, we first need to initialize our context. Notice that we will use the `JavaSparkContext` class here instead of the `SparkContext` class that we used earlier. We will use the `JavaSparkContext` class in the same way to access our data using `textFile` and then split each row into the required fields. Note how we used an anonymous class to define a split function that performs the string processing, in the highlighted code:

```java
JavaSparkContext sc = new JavaSparkContext("local[2]",
"First Spark App");
// we take the raw data in CSV format and convert it into a
set of records of the form (user, product, price)
JavaRDD<String[]> data =
sc.textFile("data/UserPurchaseHistory.csv")
.map(new Function<String, String[]>() {
  @Override
  public String[] call(String s) throws Exception {
    return s.split(",");
  }
});
```

Now, we can compute the same metrics as we did in our Scala example. Note how some methods are the same (for example, `distinct` and `count`) for the Java and Scala APIs. Also note the use of anonymous classes that we pass to the `map` function. This code is highlighted here:

```java
// let's count the number of purchases
long numPurchases = data.count();
// let's count how many unique users made purchases
long uniqueUsers = data.map(new Function<String[], String>() {
  @Override
  public String call(String[] strings) throws Exception {
    return strings[0];
  }
}).distinct().count();
// let's sum up our total revenue
double totalRevenue = data.map(new DoubleFunction<String[]>() {
  @Override
  public Double call(String[] strings) throws Exception {
    return Double.parseDouble(strings[2]);
  }
}).sum();
```

In the following lines of code, we can see that the approach to compute the most popular product is the same as that in the Scala example. The extra code might seem complex, but it is mostly related to the Java code required to create the anonymous functions (which we have highlighted here). The actual functionality is the same:

```java
// let's find our most popular product
// first we map the data to records of (product, 1) using a PairFunction
// and the Tuple2 class.
// then we call a reduceByKey operation with a Function2,
which is essentially the sum function
List<Tuple2<String, Integer>> pairs = data.map(new
PairFunction<String[], String, Integer>() {
  @Override
  public Tuple2<String, Integer> call(String[] strings)
  throws Exception {
    return new Tuple2(strings[1], 1);
  }
}).reduceByKey(new Function2<Integer, Integer, Integer>() {
  @Override
  public Integer call(Integer integer, Integer integer2)
  throws Exception {
    return integer + integer2;
  }
}).collect();
// finally we sort the result. Note we need to create a Comparator function,
// that reverses the sort order.
Collections.sort(pairs, new Comparator<Tuple2<String,
Integer>>() {
  @Override
  public int compare(Tuple2<String, Integer> o1,
  Tuple2<String, Integer> o2) {
    return -(o1._2() - o2._2());
  }
});
String mostPopular = pairs.get(0)._1();
int purchases = pairs.get(0)._2();
System.out.println("Total purchases: " + numPurchases);
System.out.println("Unique users: " + uniqueUsers);
System.out.println("Total revenue: " + totalRevenue);
System.out.println(String.format("Most popular product:
%s with %d purchases", mostPopular, purchases));
  }
}
```

Getting Up and Running with Spark

As can be seen, the general structure is similar to the Scala version, apart from the extra boilerplate code to declare variables and functions via anonymous inner classes. It is a good exercise to work through both examples and compare the same lines of Scala code to those in Java to understand how the same result is achieved in each language.

This program can be run with the following command executed from the project's base directory:

```
>mvn exec:java -Dexec.mainClass="JavaApp"
```

You will see output that looks very similar to the Scala version, with the results of the computation identical:

```
...
14/01/30 17:02:43 INFO spark.SparkContext: Job finished: collect at
JavaApp.java:46, took 0.039167 s
Total purchases: 5
Unique users: 4
Total revenue: 39.91
Most popular product: iPhone Cover with 2 purchases
```

The first step to a Spark program in Python

Spark's Python API exposes virtually all the functionalities of Spark's Scala API in the Python language. There are some features that are not yet supported (for example, graph processing with GraphX and a few API methods here and there). See the Python section of the *Spark Programming Guide* (http://spark.apache.org/docs/latest/programming-guide.html) for more details.

Following on from the preceding examples, we will now write a Python version. We assume that you have Python version 2.6 and higher installed on your system (for example, most Linux and Mac OS X systems come with Python preinstalled).

The example program is included in the sample code for this chapter, in the directory named `python-spark-app`, which also contains the CSV data file under the `data` subdirectory. The project contains a script, `pythonapp.py`, provided here:

```
"""A simple Spark app in Python"""
from pyspark import SparkContext
```

```python
sc = SparkContext("local[2]", "First Spark App")
# we take the raw data in CSV format and convert it into a set of
records of the form (user, product, price)
data = sc.textFile("data/UserPurchaseHistory.csv").map(lambda line:
line.split(",")).map(lambda record: (record[0], record[1], record[2]))
# let's count the number of purchases
numPurchases = data.count()
# let's count how many unique users made purchases
uniqueUsers = data.map(lambda record: record[0]).distinct().count()
# let's sum up our total revenue
totalRevenue = data.map(lambda record: float(record[2])).sum()
# let's find our most popular product
products = data.map(lambda record: (record[1], 1.0)).
reduceByKey(lambda a, b: a + b).collect()
mostPopular = sorted(products, key=lambda x: x[1], reverse=True)[0]

print "Total purchases: %d" % numPurchases
print "Unique users: %d" % uniqueUsers
print "Total revenue: %2.2f" % totalRevenue
print "Most popular product: %s with %d purchases" % (mostPopular[0],
mostPopular[1])
```

If you compare the Scala and Python versions of our program, you will see that generally, the syntax looks very similar. One key difference is how we express anonymous functions (also called `lambda` functions; hence, the use of this keyword for the Python syntax). In Scala, we've seen that an anonymous function mapping an input x to an output y is expressed as x => y, while in Python, it is `lambda x: y`. In the highlighted line in the preceding code, we are applying an anonymous function that maps two inputs, a and b, generally of the same type, to an output. In this case, the function that we apply is the *plus* function; hence, `lambda a, b: a + b`.

The best way to run the script is to run the following command from the base directory of the sample project:

>$SPARK_HOME/bin/spark-submit pythonapp.py

Here, the `SPARK_HOME` variable should be replaced with the path of the directory in which you originally unpacked the Spark prebuilt binary package at the start of this chapter.

Upon running the script, you should see output similar to that of the Scala and Java examples, with the results of our computation being the same:

...

14/01/30 11:43:47 INFO SparkContext: Job finished: collect at pythonapp. py:14, took 0.050251 s

Total purchases: 5

```
Unique users: 4
Total revenue: 39.91
Most popular product: iPhone Cover with 2 purchases
```

Getting Spark running on Amazon EC2

The Spark project provides scripts to run a Spark cluster in the cloud on Amazon's EC2 service. These scripts are located in the `ec2` directory. You can run the `spark-ec2` script contained in this directory with the following command:

```
>./ec2/spark-ec2
```

Running it in this way without an argument will show the help output:

```
Usage: spark-ec2 [options] <action> <cluster_name>
<action> can be: launch, destroy, login, stop, start, get-master

Options:
...
```

Before creating a Spark EC2 cluster, you will need to ensure you have an Amazon account.

> If you don't have an Amazon Web Services account, you can sign up at `http://aws.amazon.com/`.
> The AWS console is available at `http://aws.amazon.com/console/`.

You will also need to create an Amazon EC2 key pair and retrieve the relevant security credentials. The Spark documentation for EC2 (available at `http://spark.apache.org/docs/latest/ec2-scripts.html`) explains the requirements:

> *Create an Amazon EC2 key pair for yourself. This can be done by logging into your Amazon Web Services account through the AWS console, clicking on **Key Pairs** on the left sidebar, and creating and downloading a key. Make sure that you set the permissions for the private key file to 600 (that is, only you can read and write it) so that* `ssh` *will work.*
>
> *Whenever you want to use the* `spark-ec2` *script, set the environment variables* `AWS_ACCESS_KEY_ID` *and* `AWS_SECRET_ACCESS_KEY` *to your Amazon EC2 access key ID and secret access key, respectively. These can be obtained from the AWS homepage by clicking* **Account** | **Security Credentials** | **Access Credentials**.

When creating a key pair, choose a name that is easy to remember. We will simply use spark for the key pair name. The key pair file itself will be called spark.pem. As mentioned earlier, ensure that the key pair file permissions are set appropriately and that the environment variables for the AWS credentials are exported using the following commands:

```
>chmod 600 spark.pem
>export AWS_ACCESS_KEY_ID="..."
>export AWS_SECRET_ACCESS_KEY="..."
```

You should also be careful to keep your downloaded key pair file safe and not lose it, as it can only be downloaded once when it is created!

Note that launching an Amazon EC2 cluster in the following section will *incur costs* to your AWS account.

Launching an EC2 Spark cluster

We're now ready to launch a small Spark cluster by changing into the ec2 directory and then running the cluster launch command:

```
>cd ec2
>./spark-ec2 -k spark -i spark.pem -s 1 --instance-type m3.medium
--hadoop-major-version 2 launch test-cluster
```

This will launch a new Spark cluster called test-cluster with one master and one slave node of instance type m3.medium. This cluster will be launched with a Spark version built for Hadoop 2. The key pair name we used is spark, and the key pair file is spark.pem (if you gave the files different names or have an existing AWS key pair, use that name instead).

It might take quite a while for the cluster to fully launch and initialize. You should see something like this screenshot immediately after running the launch command:

```
Setting up security groups...
Creating security group test-cluster-master
Creating security group test-cluster-slaves
Searching for existing cluster test-cluster...
Spark AMI: ami-35b1885c
Launching instances...
Launched 1 slaves in us-east-1c, regid = r-5f328e75
Launched master in us-east-1c, regid = r-c0308cea
Waiting for instances to start up...
Waiting 120 more seconds...
```

Getting Up and Running with Spark

If the cluster has launched successfully, you should eventually see the console output similar to the following screenshot:

```
ec2-54-91-61-225.compute-1.amazonaws.com: Killed 0 processes
Starting master @ ec2-54-227-127-14.compute-1.amazonaws.com
ec2-54-91-61-225.compute-1.amazonaws.com: TACHYON_LOGS_DIR: /root/tachyon/libexec/../logs
ec2-54-91-61-225.compute-1.amazonaws.com: Formatting RamFS: /mnt/ramdisk (2470mb)
ec2-54-91-61-225.compute-1.amazonaws.com: Starting worker @ ip-10-182-117-29.ec2.internal
Setting up ganglia
RSYNC'ing /etc/ganglia to slaves...
ec2-54-91-61-225.compute-1.amazonaws.com
Shutting down GANGLIA gmond:                               [FAILED]
Starting GANGLIA gmond:                                    [  OK  ]
Shutting down GANGLIA gmond:                               [FAILED]
Starting GANGLIA gmond:                                    [  OK  ]
Connection to ec2-54-91-61-225.compute-1.amazonaws.com closed.
Shutting down GANGLIA gmetad:                              [FAILED]
Starting GANGLIA gmetad:                                   [  OK  ]
Stopping httpd:                                            [FAILED]
Starting httpd: httpd: Syntax error on line 153 of /etc/httpd/conf/httpd.conf: Cannot load modules/mod_authn_alias.so int
o server: /etc/httpd/modules/mod_authn_alias.so: cannot open shared object file: No such file or directory
                                                           [FAILED]
Connection to ec2-54-227-127-14.compute-1.amazonaws.com closed.
Spark standalone cluster started at http://ec2-54-227-127-14.compute-1.amazonaws.com:8080
Ganglia started at http://ec2-54-227-127-14.compute-1.amazonaws.com:5080/ganglia
Done!
Nicks-MacBook-Pro:ec2 Nick$
```

To test whether we can connect to our new cluster, we can run the following command:

```
>ssh -i spark.pem root@ec2-54-227-127-14.compute-1.amazonaws.com
```

Remember to replace the public domain name of the master node (the address after `root@` in the preceding command) with the correct Amazon EC2 public domain name that will be shown in your console output after launching the cluster.

You can also retrieve your cluster's master public domain name by running this line of code:

```
>./spark-ec2 -i spark.pem get-master test-cluster
```

After successfully running the `ssh` command, you will be connected to your Spark master node in EC2, and your terminal output should match the following screenshot:

```
       _|  _|_  )
       _|  (    /    Amazon Linux AMI
       _|\___|___|

https://aws.amazon.com/amazon-linux-ami/2013.03-release-notes/
There are 60 security update(s) out of 254 total update(s) available
Run "sudo yum update" to apply all updates.
Amazon Linux version 2014.09 is available.
root@ip-10-150-79-53 ~]$
```

Chapter 1

We can test whether our cluster is correctly set up with Spark by changing into the Spark directory and running an example in the local mode:

```
>cd spark
>MASTER=local[2] ./bin/run-example SparkPi
```

You should see output similar to running the same command on your local computer:

```
...
14/01/30 20:20:21 INFO SparkContext: Job finished: reduce at SparkPi.
scala:35, took 0.864044012 s
Pi is roughly 3.14032
...
```

Now that we have an actual cluster with multiple nodes, we can test Spark in the cluster mode. We can run the same example on the cluster, using our 1 slave node, by passing in the master URL instead of the local version:

```
>MASTER=spark://ec2-54-227-127-14.compute-1.amazonaws.com:7077 ./bin/run-example SparkPi
```

> Note that you will need to substitute the preceding master domain name with the correct domain name for your specific cluster.

Again, the output should be similar to running the example locally; however, the log messages will show that your driver program has connected to the Spark master:

```
...
14/01/30 20:26:17 INFO client.Client$ClientActor: Connecting to master spark://ec2-54-220-189-136.eu-west-1.compute.amazonaws.com:7077
14/01/30 20:26:17 INFO cluster.SparkDeploySchedulerBackend: Connected to Spark cluster with app ID app-20140130202617-0001
14/01/30 20:26:17 INFO client.Client$ClientActor: Executor added: app-20140130202617-0001/0 on worker-20140130201049-ip-10-34-137-45.eu-west-1.compute.internal-57119 (ip-10-34-137-45.eu-west-1.compute.internal:57119) with 1 cores
14/01/30 20:26:17 INFO cluster.SparkDeploySchedulerBackend: Granted executor ID app-20140130202617-0001/0 on hostPort ip-10-34-137-45.eu-west-1.compute.internal:57119 with 1 cores, 2.4 GB RAM
```

Getting Up and Running with Spark

```
14/01/30 20:26:17 INFO client.Client$ClientActor: Executor updated: app-
20140130202617-0001/0 is now RUNNING
14/01/30 20:26:18 INFO spark.SparkContext: Starting job: reduce at
SparkPi.scala:39
...
```

Feel free to experiment with your cluster. Try out the interactive console in Scala, for example:

```
> ./bin/spark-shell --master spark://ec2-54-227-127-14.compute-1.
amazonaws.com:7077
```

Once you've finished, type `exit` to leave the console. You can also try the PySpark console by running the following command:

```
> ./bin/pyspark --master spark://ec2-54-227-127-14.compute-1.amazonaws.
com:7077
```

You can use the Spark Master web interface to see the applications registered with the master. To load the Master Web UI, navigate to `ec2-54-227-127-14.compute-1.amazonaws.com:8080` (again, remember to replace this domain name with your own master domain name). You should see something similar to the following screenshot showing the example you ran as well as the two console applications you launched:

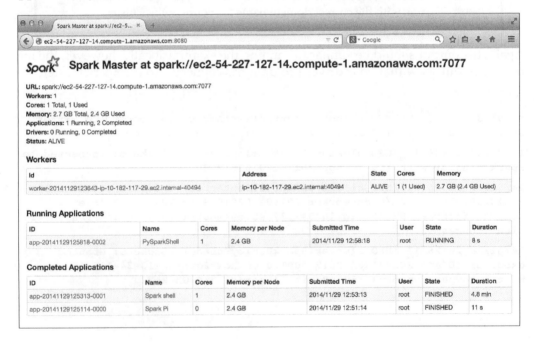

Remember that *you will be charged by Amazon* for usage of the cluster. Don't forget to stop or terminate this test cluster once you're done with it. To do this, you can first exit the `ssh` session by typing `exit` to return to your own local system and then, run the following command:

```
>./ec2/spark-ec2 -k spark -i spark.pem destroy test-cluster
```

You should see the following output:

```
Are you sure you want to destroy the cluster test-cluster?
The following instances will be terminated:
Searching for existing cluster test-cluster...
Found 1 master(s), 1 slaves
> ec2-54-227-127-14.compute-1.amazonaws.com
> ec2-54-91-61-225.compute-1.amazonaws.com
ALL DATA ON ALL NODES WILL BE LOST!!
Destroy cluster test-cluster (y/N): y
Terminating master...
Terminating slaves...
```

Hit *Y* and then *Enter* to destroy the cluster.

Congratulations! You've just set up a Spark cluster in the cloud, run a fully parallel example program on this cluster, and terminated it. If you would like to try out any of the example code in the subsequent chapters (or your own Spark programs) on a cluster, feel free to experiment with the Spark EC2 scripts and launch a cluster of your chosen size and instance profile (just be mindful of the costs and remember to shut it down when you're done!).

Summary

In this chapter, we covered how to set up Spark locally on our own computer as well as in the cloud as a cluster running on Amazon EC2. You learned the basics of Spark's programming model and API using the interactive Scala console, and we wrote the same basic Spark program in Scala, Java, and Python.

In the next chapter, we will consider how to go about using Spark to create a machine learning system.

2
Designing a Machine Learning System

In this chapter, we will design a high-level architecture for an intelligent, distributed machine learning system that uses Spark as its core computation engine. The problem we will focus on will be taking the existing architecture for a web-based business and redesigning it to use automated machine learning systems to power key areas of the business. In this chapter, we will:

- Introduce our hypothetical business scenario
- Provide an overview of the current architecture
- Explore various ways in which machine learning systems can enhance or replace certain business functions
- Provide a new architecture based on these ideas

A modern large-scale data environment includes the following requirements:

- It must integrate with other components of the system, especially with data collection and storage systems, analytics and reporting, and frontend applications.
- It should be easily scalable and independent of the rest of the architecture. Ideally, this should be in the form of horizontal as well as vertical scalability.
- It should allow efficient computation in respect of the type of workload in mind, that is machine learning and iterative analytics applications.
- If possible, it should support both batch and real-time workloads.

As a framework, Spark meets these criteria. However, we must ensure that the machine learning systems designed on Spark also meet these criteria. There is no good in implementing an algorithm that ends up having bottlenecks that cause our system to fail in terms of one or more of these requirements.

Introducing MovieStream

To better illustrate the design of our architecture, we will introduce a practical scenario. Let's assume that we have just been appointed to head the data science team of MovieStream, a fictitious Internet business that streams movies and television shows to its users.

MovieStream is growing rapidly, adding both users and titles to its catalogue. The current MovieStream system is outlined in the following diagram:

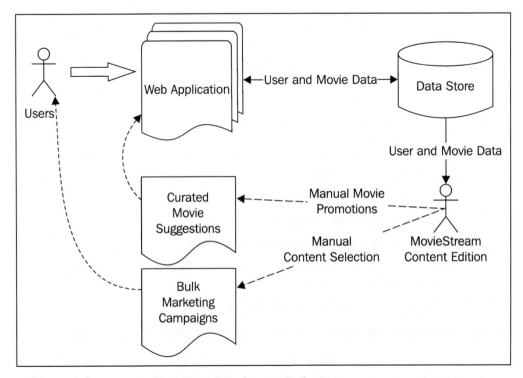

MovieStream's current architecture

As we can see in the preceding diagram, currently, MovieStream's content editorial team is responsible for deciding which movies and shows are promoted and shown on the various parts of the site. They are also responsible for creating the content for MovieStream's bulk marketing campaigns, which include e-mail and other direct marketing channels. Currently, MovieStream collects basic data on what titles are viewed by users on an aggregate basis and has access to some demographic data collected from users when they sign up to the service. In addition, they have access to some basic metadata about the titles in their catalogue.

The MovieStream team is stretched thin due to their rapid growth, and they can't keep up with the number of new releases and the growing activity of their users. The CEO of MovieStream has heard a lot about big data, machine learning, and artificial intelligence, and would like us to build a machine learning system for MovieStream that can handle many of the functions currently handled by the content team in an automated manner.

Business use cases for a machine learning system

Perhaps the first question we should answer is, "Why use machine learning at all?" Why doesn't MovieStream simply continue with human-driven decisions? There are many reasons to use machine learning (and certainly some reasons not to), but the most important ones are mentioned here:

- The scale of data involved means that full human involvement quickly becomes infeasible as MovieStream grows
- Model-driven approaches such as machine learning and statistics can often benefit from uncovering patterns that cannot be seen by humans (due to the size and complexity of the datasets)
- Model-driven approaches can avoid human and emotional biases (as long as the correct processes are carefully applied)

However, there is no reason why both model-driven and human-driven processes and decision making cannot coexist. For example, many machine learning systems rely on receiving labeled data in order to train models. Often, labeling such data is costly, time consuming, and requires human input. A good example of this is classifying textual data into categories or assigning a sentiment indicator to the text. Many real-world systems use some form of human-driven system to generate labels for such data (or at least part of it) to provide training data to models. These models are then used to make predictions in the live system at a larger scale.

In the context of MovieStream, we need not fear that our machine learning system will make the content team redundant. Indeed, we will see that our aim is to lift the burden of time-consuming tasks where machine learning might be able to perform better while providing tools to allow the team to better understand the users and content. This might, for example, help them in selecting which new content to acquire for the catalogue (which involves a significant amount of cost and is therefore a critical aspect of the business).

Personalization

Perhaps one of the most important potential applications of machine learning in MovieStream's business is personalization. Generally speaking, personalization refers to adapting the experience of a user and the content presented to them based on various factors, which might include user behavior data as well as external factors.

Recommendations are essentially a subset of personalization. Recommendation generally refers to presenting a user with a list of items that we hope the user will be interested in. Recommendations might be used in web pages (for example, recommending related products), via e-mails or other direct marketing channels, via mobile apps, and so on.

Personalization is very similar to recommendations, but while recommendations are usually focused on an *explicit* presentation of products or content to the user, personalization is more generic and, often, more *implicit*. For example, applying personalization to search on the MovieStream site might allow us to adapt the search results for a given user, based on the data available about that user. This might include recommendation-based data (in the case of a search for products or content) but might also include various other factors such as geolocation and past search history. It might not be apparent to the user that the search results are adapted to their specific profile; this is why personalization tends to be more implicit.

Targeted marketing and customer segmentation

In a manner similar to recommendations, targeted marketing uses a model to select what to target at users. While generally recommendations and personalization are focused on a one-to-one situation, segmentation approaches might try to assign users into groups based on characteristics and, possibly, behavioral data. The approach might be fairly simple or might involve a machine learning model such as clustering. Either way, the result is a set of segment assignments that might allow us to understand the broad characteristics of each group of users, what makes them similar to each other within a group, and what makes them different from others in different groups.

This could help MovieStream to better understand the drivers of user behavior and might also allow a broader targeting approach where groups are targeted as opposed to (or more likely, in addition to) direct one-to-one targeting with personalization.

These methods can also help when we don't necessarily have labeled data available (as is the case with certain user and content profile data) but we still wish to perform more focused targeting than a complete *one-size-fits-all* approach.

Predictive modeling and analytics

A third area where machine learning can be applied is in predictive analytics. This is a very broad term, and in some ways, it encompasses recommendations, personalization, and targeting too. In this context, since recommendations and segmentation are somewhat distinct, we use the term **predictive modeling** to refer to other models that seek to make predictions. An example of this can be a model to predict the potential viewing activity and revenue of new titles before any data is available on how popular the title might be. MovieStream can use past activity and revenue data, together with content attributes, to create a **regression model** that can be used to make predictions for brand new titles.

As another example, we can use a **classification model** to automatically assign tags, keywords, or categories to new titles for which we only have partial data.

Types of machine learning models

While we have highlighted a few use cases for machine learning in the context of the preceding MovieStream example, there are many other examples, some of which we will touch on in the relevant chapters when we introduce each machine learning task.

However, we can broadly divide the preceding use cases and methods into two categories of machine learning:

- **Supervised learning**: These types of models use *labeled* data to learn. Recommendation engines, regression, and classification are examples of supervised learning methods. The labels in these models can be user-movie ratings (for recommendation), movie tags (in the case of the preceding classification example), or revenue figures (for regression). We will cover supervised learning models in *Chapter 4, Building a Recommendation Engine with Spark*, *Chapter 5, Building a Classification Model with Spark*, and *Chapter 6, Building a Regression Model with Spark*.

- **Unsupervised learning**: When a model does not require labeled data, we refer to unsupervised learning. These types of models try to learn or extract some underlying structure in the data or reduce the data down to its most important features. Clustering, dimensionality reduction, and some forms of feature extraction, such as text processing, are all unsupervised techniques and will be dealt with in *Chapter 7, Building a Clustering Model with Spark*, *Chapter 8, Dimensionality Reduction with Spark*, and *Chapter 9, Advanced Text Processing with Spark*.

The components of a data-driven machine learning system

The high-level components of our machine learning system are outlined in the following diagram. This diagram illustrates the machine learning pipeline from which we obtain data and in which we store data. We then transform it into a form that is usable as input to a machine learning model; train, test, and refine our model; and then, deploy the final model to our production system. The process is then repeated as new data is generated.

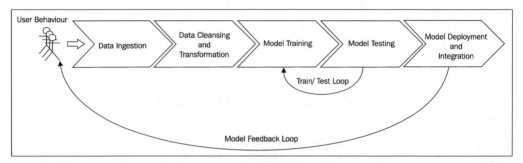

A general machine learning pipeline

Data ingestion and storage

The first step in our machine learning pipeline will be taking in the data that we require for training our models. Like many other businesses, MovieStream's data is typically generated by user activity, other systems (this is commonly referred to as machine-generated data), and external sources (for example, the time of day and weather during a particular user's visit to the site).

This data can be ingested in various ways, for example, gathering user activity data from browser and mobile application event logs or accessing external web APIs to collect data on geolocation or weather.

Once the collection mechanisms are in place, the data usually needs to be stored. This includes the raw data, data resulting from intermediate processing, and final model results to be used in production.

Data storage can be complex and involve a wide variety of systems, including HDFS, Amazon S3, and other filesystems; SQL databases such as MySQL or PostgreSQL; distributed NoSQL data stores such as HBase, Cassandra, and DynamoDB; and search engines such as Solr or Elasticsearch to stream data systems such as Kafka, Flume, or Amazon Kinesis.

For the purposes of this book, we will assume that the relevant data is available to us, so we will focus on the processing and modeling steps in the following pipeline.

Data cleansing and transformation

The majority of machine learning models operate on features, which are typically numerical representations of the input variables that will be used for the model.

While we might want to spend the majority of our time exploring machine learning models, data collected via various systems and sources in the preceding ingestion step is, in most cases, in a raw form. For example, we might log user events such as details of when a user views the information page for a movie, when they watch a movie, or when they provide some other feedback. We might also collect external information such as the location of the user (as provided through their IP address, for example). These event logs will typically contain some combination of textual and numeric information about the event (and also, perhaps, other forms of data such as images or audio).

In order to use this raw data in our models, in almost all cases, we need to perform preprocessing, which might include:

- **Filtering data**: Let's assume that we want to create a model from a subset of the raw data, such as only the most recent few months of activity data or only events that match certain criteria.

- **Dealing with missing, incomplete, or corrupted data**: Many real-world datasets are incomplete in some way. This might include data that is missing (for example, due to a missing user input) or data that is incorrect or flawed (for example, due to an error in data ingestion or storage, technical issues or bugs, or software or hardware failure). We might need to filter out bad data or alternatively decide a method to fill in missing data points (such as using the average value from the dataset for missing points, for example).

- **Dealing with potential anomalies, errors, and outliers**: Erroneous or outlier data might skew the results of model training, so we might wish to filter these cases out or use techniques that are able to deal with outliers.

- **Joining together disparate data sources**: For example, we might need to match up the event data for each user with different internal data sources, such as user profiles, as well as external data, such as geolocation, weather, and economic data.
- **Aggregating data**: Certain models might require input data that is aggregated in some way, such as computing the sum of a number of different event types per user.

Once we have performed initial preprocessing on our data, we often need to transform the data into a representation that is suitable for machine learning models. For many model types, this representation will take the form of a vector or matrix structure that contains numerical data. Common challenges during data transformation and feature extraction include:

- Taking categorical data (such as country for geolocation or category for a movie) and encoding it in a numerical representation.
- Extracting useful features from text data.
- Dealing with image or audio data.
- We often convert numerical data into categorical data to reduce the number of values a variable can take on. An example of this is converting a variable for age into buckets (such as 25-35, 45-55, and so on).
- Transforming numerical features; for example, applying a log transformation to a numerical variable can help deal with variables that take on a very large range of values.
- Normalizing and standardizing numerical features ensures that all the different input variables for a model have a consistent scale. Many machine learning models require standardized input to work properly.
- Feature engineering is the process of combining or transforming the existing variables to create new features. For example, we can create a new variable that is the average of some other data, such as the average number of times a user watches a movie.

We will cover all of these techniques through the examples in this book.

These data-cleansing, exploration, aggregation, and transformation steps can be carried out using both Spark's core API functions as well as the SparkSQL engine, not to mention other external Scala, Java, or Python libraries. We can take advantage of Spark's Hadoop compatibility to read data from and write data to the various different storage systems mentioned earlier.

Model training and testing loop

Once we have our training data in a form that is suitable for our model, we can proceed with the model's training and testing phase. During this phase, we are primarily concerned with **model selection**. This can refer to choosing the best modeling approach for our task, or the best parameter settings for a given model. In fact, the term model selection often refers to both of these processes, as, in many cases, we might wish to try out various models and select the best performing model (with the best performing parameter settings for each model). It is also common to explore the application of combinations of different models (known as **ensemble methods**) in this phase.

This is typically a fairly straightforward process of running our chosen model on our training dataset and testing its performance on a test dataset (that is, a set of data that is held out for the evaluation of the model that the model has not seen in the training phase). This process is referred to as **cross-validation**.

However, due to the large scale of data we are typically working with, it is often useful to carry out this initial train-test loop on a smaller representative sample of our full dataset or perform model selection using parallel methods where possible.

For this part of the pipeline, Spark's built-in machine learning library, MLlib, is a perfect fit. We will focus most of our attention in this book on the model training, evaluation, and cross-validation steps for various machine learning techniques, using MLlib and Spark's core features.

Model deployment and integration

Once we have found the optimal model based on the train-test loop, we might still face the task of deploying the model to a production system so that it can be used to make actionable predictions.

Usually, this process involves exporting the trained model to a central data store from where the production-serving system can obtain the latest version. Thus, the live system *refreshes* the model periodically as a new model is trained.

Model monitoring and feedback

It is critically important to monitor the performance of our machine learning system in production. Once we deploy our optimal trained model, we wish to understand how it is doing in the "wild". Is it performing as we expect on new, unseen data? Is its accuracy good enough? The reality is regardless of how much model selection and tuning we try to do in the earlier phases; the only way to measure true performance is to observe what happens in our production system.

Also, bear in mind that model accuracy and predictive performance is only one aspect of a real-world system. Usually, we are concerned with other metrics related to business performance (for example, revenue and profitability) or user experience (such as the time spent on our site and how active our users are overall). In most cases, we cannot easily map model-predictive performance to these business metrics. The accuracy of a recommendation or targeting system might be important, but it relates only indirectly to the true metrics we are concerned about, namely whether we are improving user experience, activity, and ultimately, revenue.

So, in real-world systems, we should monitor both model-accuracy metrics as well as business metrics. If possible, we should be able to experiment with different models running in production to allow us to optimize against these business metrics by making changes to the models. This is often done using live split tests. However, doing this correctly is not an easy task, and live testing and experimentation is expensive, in the sense that mistakes, poor performance, and using baseline models (they provide a control against which we test out production models) can negatively impact user experience and revenue.

Another important aspect of this phase is **model feedback**. This is the process where the predictions of our model feed through into user behavior; this, in turn, feeds through into our model. In a real-world system, our models are essentially influencing their own future training data by impacting decision-making and potential user behavior.

For example, if we have deployed a recommendation system, then, by making recommendations, we might be influencing user behavior because we are only allowing users a limited selection of choices. We hope that this selection is relevant due to our model; however, this feedback loop, in turn, can influence our model's training data. This, in turn, feeds back into real-world performance. It is possible to get into an ever-narrowing feedback loop; ultimately, this can negatively affect both model accuracy and our important business metrics.

Fortunately, there are mechanisms by which we can try to limit the potential negative impact of this feedback loop. These include providing some unbiased training data by having a small portion of data coming from users who are not exposed to our models or by being principled in the way we balance exploration, to learn more about our data, and exploitation, to use what we have learned to improve our system's performance.

We will briefly cover some aspects of real-time monitoring and model updates in *Chapter 10, Real-time Machine Learning with Spark Streaming*.

Batch versus real time

In the previous sections, we outlined the common batch processing approach, where the model is retrained using all data or a subset of all data, periodically. As the preceding pipeline takes some time to complete, it might not be possible to use this approach to update models immediately as new data arrives.

While we will be mostly covering batch machine learning approaches in this book, there is a class of machine learning algorithms known as **online learning**; they update immediately as new data is fed into the model, thus enabling a real-time system. A common example is an online-optimization algorithm for a linear model, such as stochastic gradient descent. We can learn this algorithm using examples. The advantages of these methods are that the system can react very quickly to new information and also that the system can adapt to changes in the underlying behavior (that is, if the characteristics and distribution of the input data are changing over time, which is almost always the case in real-world situations).

However, online-learning models come with their own unique challenges in a production context. For example, it might be difficult to ingest and transform data in real time. It can also be complex to properly perform model selection in a purely online setting. Latency of the online training and the model selection and deployment phases might be too high for true real-time requirements (for example, in online advertising, latency requirements are measured in single-digit milliseconds). Finally, batch-oriented frameworks might make it awkward to handle real-time processes of a streaming nature.

Fortunately, Spark's real-time stream processing component, **Spark Streaming**, is a good potential fit for real-time machine learning workflows. We will explore Spark Streaming and online learning in *Chapter 10, Real-time Machine Learning with Spark Streaming*.

Due to the complexities inherent in a true real-time machine learning system, in practice, many systems target near real-time operations. This is essentially a hybrid approach where models are not necessarily updated immediately as new data arrives; instead, the new data is collected into mini-batches of a small set of training data. These mini-batches can be fed to an online-learning algorithm. In many cases, this approach is combined with a periodic batch process that might recompute the model on the entire data set and perform more complex processing and model selection. This can help ensure that the real-time model does not degrade over time.

Another similar approach involves making approximate updates to a more complex model as new data arrives while recomputing the entire model in a batch process periodically. In this way, the model can learn from new data, with a short delay (usually measured in seconds or, perhaps, a few minutes), but will become more and more inaccurate over time due to the approximation applied. The periodic recomputation takes care of this by retraining the model on all available data.

An architecture for a machine learning system

Now that we have explored how our machine learning system might work in the context of MovieStream, we can outline a possible architecture for our system:

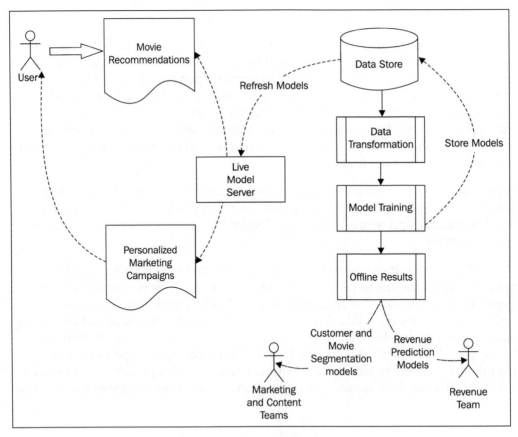

MovieStream's future architecture

As we can see, our system incorporates the machine learning pipeline outlined in the preceding diagram; this system also includes:

- Collecting data about users, their behavior, and our content titles
- Transforming this data into features
- Training our models, including our training-testing and model-selection phases
- Deploying the trained models to both our live model-serving system as well as using these models for offline processes
- Feeding back the model results into the MovieStream website through recommendation and targeting pages
- Feeding back the model results into MovieStream's personalized marketing channels
- Using the offline models to provide tools to MovieStream's various teams to better understand user behavior, characteristics of the content catalogue, and drivers of revenue for the business

Practical exercise

Imagine that you now need to provide input to the frontend and infrastructure engineering team about the data that your machine learning system will need. Consider a brief for them on how they should structure the data-collection mechanisms. Write down some examples of what the raw data might look like (for example, web logs, event logs, and so on) and how it should flow through the system. Take into account the following aspects:

- What data sources will be required
- What format should the data be in
- How often should data be collected, processed, potentially aggregated, and stored
- What data storage will you use to ensure scalability

Summary

In this chapter, you learned about the components inherent in a data-driven, automated machine learning system. We also outlined how a possible high-level architecture for such a system might look in a real-world situation.

In the next chapter, we will discuss how to obtain publicly-available datasets for common machine learning tasks. We will also explore general concepts related to processing, cleaning, and transforming data so that they can be used to train a machine learning model.

3
Obtaining, Processing, and Preparing Data with Spark

Machine learning is an extremely broad field, and these days, applications can be found across areas that include web and mobile applications, Internet of Things and sensor networks, financial services, healthcare, and various scientific fields, to name just a few.

Therefore, the range of data available for potential use in machine learning is enormous. In this book, we will focus mostly on business applications. In this context, the data available often consists of data internal to an organization (such as transactional data for a financial services company) as well as external data sources (such as financial asset price data for the same financial services company).

For example, recall from *Chapter 2, Designing a Machine Learning System,* that the main internal source of data for our hypothetical Internet business, MovieStream, consists of data on the movies available on the site, the users of the service, and their behavior. This includes data about movies and other content (for example, title, categories, description, images, actors, and directors), user information (for example, demographics, location, and so on), and user activity data (for example, web page views, title previews and views, ratings, reviews, and social data such as *likes*, *shares*, and social network profiles on services including Facebook and Twitter).

External data sources in this example might include weather and geolocation services, third-party movie ratings and review sites such as *IMDB* and *Rotten Tomatoes*, and so on.

Generally speaking, it is quite difficult to obtain data of an internal nature for real-world services and businesses, as it is commercially sensitive (in particular, data on purchasing activity, user or customer behavior, and revenue) and of great potential value to the organization concerned. This is why it is also often the most useful and interesting data on which to apply machine learning—a good machine learning model that can make accurate predictions can be highly valuable (witness the success of machine learning competitions such as the *Netflix Prize* and *Kaggle*).

In this book, we will make use of datasets that are publicly available to illustrate concepts around data processing and training of machine learning models.

In this chapter, we will:

- Briefly cover the types of data typically used in machine learning.
- Provide examples of where to obtain interesting datasets, often publicly available on the Internet. We will use some of these datasets throughout the book to illustrate the use of the models we introduce.
- Discover how to process, clean, explore, and visualize our data.
- Introduce various techniques to transform our raw data into features that can be used as input to machine learning algorithms.
- Learn how to normalize input features using external libraries as well as Spark's built-in functionality.

Accessing publicly available datasets

Fortunately, while commercially-sensitive data can be hard to come by, there are still a number of useful datasets available publicly. Many of these are often used as benchmark datasets for specific types of machine learning problems. Examples of common data sources include:

- **UCI Machine Learning Repository**: This is a collection of almost 300 datasets of various types and sizes for tasks including classification, regression, clustering, and recommender systems. The list is available at `http://archive.ics.uci.edu/ml/`.
- **Amazon AWS public datasets**: This is a set of often very large datasets that can be accessed via Amazon S3. These datasets include the Human Genome Project, the Common Crawl web corpus, Wikipedia data, and Google Books Ngrams. Information on these datasets can be found at `http://aws.amazon.com/publicdatasets/`.

- **Kaggle**: This is a collection of datasets used in machine learning competitions run by Kaggle. Areas include classification, regression, ranking, recommender systems, and image analysis. These datasets can be found under the *Competitions* section at http://www.kaggle.com/competitions.
- **KDnuggets**: This has a detailed list of public datasets, including some of those mentioned earlier. The list is available at http://www.kdnuggets.com/datasets/index.html.

> There are many other resources to find public datasets depending on the specific domain and machine learning task. Hopefully, you might also have exposure to some interesting academic or commercial data of your own!

To illustrate a few key concepts related to data processing, transformation, and feature extraction in Spark, we will download a commonly-used dataset for movie recommendations; this dataset is known as the **MovieLens** dataset. As it is applicable to recommender systems as well as potentially other machine learning tasks, it serves as a useful example dataset.

> Spark's machine learning library, MLlib, has been under heavy development since its inception, and unlike the Spark core, it is still not in a fully stable state with regard to its overall API and design.
>
> As of Spark Version 1.2.0, a new, experimental API for MLlib has been released under the ml package (whereas the current library resides under the mllib package). This new API aims to enhance the APIs and interfaces for models as well as feature extraction and transformation so as to make it easier to build pipelines that chain together steps that include feature extraction, normalization, dataset transformations, model training, and cross-validation.
>
> In the upcoming chapters, we will only cover the existing, more developed MLlib API, since the new API is still experimental and may be subject to major changes in the next few Spark releases. Over time, the various feature-processing techniques and models that we will cover will simply be ported to the new API; however, the core concepts and most underlying code will remain largely unchanged.

The MovieLens 100k dataset

The MovieLens 100k dataset is a set of 100,000 data points related to ratings given by a set of users to a set of movies. It also contains movie metadata and user profiles. While it is a small dataset, you can quickly download it and run Spark code on it. This makes it ideal for illustrative purposes.

You can download the dataset from `http://files.grouplens.org/datasets/movielens/ml-100k.zip`.

Once you have downloaded the data, unzip it using your terminal:

```
>unzip ml-100k.zip
   inflating: ml-100k/allbut.pl
   inflating: ml-100k/mku.sh
   inflating: ml-100k/README
   ...
   inflating: ml-100k/ub.base
   inflating: ml-100k/ub.test
```

This will create a directory called `ml-100k`. Change into this directory and examine the contents. The important files are `u.user` (user profiles), `u.item` (movie metadata), and `u.data` (the ratings given by users to movies):

```
>cd ml-100k
```

The `README` file contains more information on the dataset, including the variables present in each data file. We can use the `head` command to examine the contents of the various files.

For example, we can see that the `u.user` file contains the `user id`, `age`, `gender`, `occupation`, and `ZIP code` fields, separated by a pipe (`|` character):

```
>head -5 u.user
   1|24|M|technician|85711
   2|53|F|other|94043
   3|23|M|writer|32067
   4|24|M|technician|43537
   5|33|F|other|15213
```

The u.item file contains the movie id, title, release data, and IMDB link fields and a set of fields related to movie category data. It is also separated by a | character:

```
>head -5 u.item
 1|Toy Story (1995)|01-Jan-1995||http://us.imdb.com/M/title-exact?Toy%20
Story%20(1995)|0|0|0|1|1|1|0|0|0|0|0|0|0|0|0|0|0|0|0
 2|GoldenEye (1995)|01-Jan-1995||http://us.imdb.com/M/title-
exact?GoldenEye%20(1995)|0|1|1|0|0|0|0|0|0|0|0|0|0|0|0|0|1|0|0
 3|Four Rooms (1995)|01-Jan-1995||http://us.imdb.com/M/title-
exact?Four%20Rooms%20(1995)|0|0|0|0|0|0|0|0|0|0|0|0|0|0|0|0|1|0|0
 4|Get Shorty (1995)|01-Jan-1995||http://us.imdb.com/M/title-
exact?Get%20Shorty%20(1995)|0|1|0|0|0|1|0|0|1|0|0|0|0|0|0|0|0|0|0
 5|Copycat (1995)|01-Jan-1995||http://us.imdb.com/M/title-
exact?Copycat%20(1995)|0|0|0|0|0|0|1|0|1|0|0|0|0|0|0|0|1|0|0
```

Finally, the u.data file contains the user id, movie id, rating (1-5 scale), and timestamp fields and is separated by a tab (the \t character):

```
>head -5 u.data
196     242     3       881250949
186     302     3       891717742
22      377     1       878887116
244     51      2       880606923
166     346     1       886397596
```

Exploring and visualizing your data

Now that we have our data available, let's fire up an interactive Spark console and explore it! For this section, we will use Python and the PySpark shell, as we are going to use the IPython interactive console and the matplotlib plotting library to process and visualize our data.

> IPython is an advanced, interactive shell for Python. It includes a useful set of features called pylab, which includes NumPy and SciPy for numerical computing and matplotlib for interactive plotting and visualization.
>
> We recommend that you use the latest version of IPython (2.3.1 at the time of writing this book). To install IPython for your platform, follow the instructions available at http://ipython.org/install.html. If this is the first time you are using IPython, you can find a tutorial at http://ipython.org/ipython-doc/stable/interactive/tutorial.html.

Obtaining, Processing, and Preparing Data with Spark

You will need to install all the packages listed earlier in order to work through the code in this chapter. Instructions to install the packages can be found in the code bundle. If you are starting out with Python or are unfamiliar with the process of installing these packages, we strongly recommend that you use a prebuilt scientific Python installation such as Anaconda (available at `http://continuum.io/downloads`) or Enthought (available at `https://store.enthought.com/downloads/`). These make the installation process much easier and include everything you will need to follow the example code.

The PySpark console allows the option of setting which Python executable needs to be used to run the shell. We can choose to use IPython, as opposed to the standard Python shell, when launching our PySpark console. We can also pass in additional options to IPython, including telling it to launch with the pylab functionality enabled.

We can do this by running the following command from the Spark home directory (that is, the same directory that we used previously to explore the Spark interactive console):

```
>IPYTHON=1 IPYTHON_OPTS="--pylab" ./bin/pyspark
```

You will see the PySpark console start up, showing output similar to the following screenshot:

The PySpark console using IPython

Chapter 3

> Notice the IPython 2.3.1 -- An enhanced Interactive Python and Using matplotlib backend: MacOSX lines; they indicate that both the IPython and pylab functionalities are being used by the PySpark shell.
>
> You might see a slightly different output, depending on your operating system and software versions.

Now that we have our IPython console open, we can start to explore the MovieLens dataset and do some basic analysis.

> You can follow along with this chapter by entering the code examples into your IPython console. IPython also provides an HTML-enabled Notebook application. It provides some enhanced functionality over the standard IPython console, such as inline graphics for plotting, the HTML markup functionality, as well as the ability to run cells of code independently.
>
> The images used in this chapter were generated using the IPython Notebook, so don't worry if yours look a little bit different in style, as long as they contain the same content! You can also use the Notebook for the code in this chapter, if you prefer. In addition to the Python code for this chapter, we have provided a version saved in the IPython Notebook format, which you can load into your own IPython Notebook.
>
> Check out the instructions on how to use the IPython Notebook at http://ipython.org/ipython-doc/stable/interactive/notebook.html.

Exploring the user dataset

First, we will analyze the characteristics of MovieLens users. Enter the following lines into your console (where PATH refers to the base directory in which you performed the unzip command to unzip the preceding MovieLens 100k dataset):

```
user_data = sc.textFile("/PATH/ml-100k/u.user")
user_data.first()
```

You should see output similar to this:

`u'1|24|M|technician|85711'`

As we can see, this is the first line of our user data file, separated by the " | " character.

 The first function is similar to collect, but it only returns the first element of the RDD to the driver. We can also use take(k) to collect only the first *k* elements of the RDD to the driver.

Let's transform the data by splitting each line, around the " | " character. This will give us an RDD where each record is a Python list that contains the user ID, age, gender, occupation, and ZIP code fields.

We will then count the number of users, genders, occupations, and ZIP codes. We can achieve this by running the following code in the console, line by line. Note that we do not cache the data, as it is unnecessary for this small size:

```
user_fields = user_data.map(lambda line: line.split("|"))
num_users = user_fields.map(lambda fields: fields[0]).count()
num_genders = user_fields.map(lambda fields: fields[2]).distinct().count()
num_occupations = user_fields.map(lambda fields: fields[3]).distinct().count()
num_zipcodes = user_fields.map(lambda fields: fields[4]).distinct().count()
print "Users: %d, genders: %d, occupations: %d, ZIP codes: %d" % (num_users, num_genders, num_occupations, num_zipcodes)
```

You will see the following output:

Users: 943, genders: 2, occupations: 21, ZIP codes: 795

Next, we will create a histogram to analyze the distribution of user ages, using matplotlib's hist function:

```
ages = user_fields.map(lambda x: int(x[1])).collect()
hist(ages, bins=20, color='lightblue', normed=True)
fig = matplotlib.pyplot.gcf()
fig.set_size_inches(16, 10)
```

We passed in the ages array, together with the number of bins for our histogram (20 in this case), to the hist function. Using the normed=True argument, we also specified that we want the histogram to be normalized so that each bucket represents the percentage of the overall data that falls into that bucket.

You will see an image containing the histogram chart, which looks something like the one shown here. As we can see, the ages of MovieLens users are somewhat skewed towards younger viewers. A large number of users are between the ages of about 15 and 35.

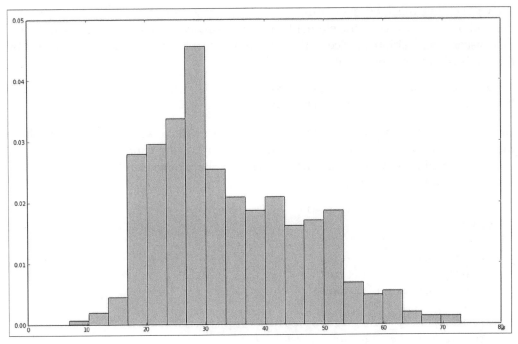

Distribution of user ages

We might also want to explore the relative frequencies of the various occupations of our users. We can do this using the following code snippet. First, we will use the MapReduce approach introduced previously to count the occurrences of each occupation in the dataset. Then, we will use matplotlib to display a bar chart of occupation counts, using the bar function.

Since part of our data is the descriptions of textual occupation, we will need to manipulate it a little to get it to work with the bar function:

```
count_by_occupation = user_fields.map(lambda fields: (fields[3], 1)).
reduceByKey(lambda x, y: x + y).collect()
x_axis1 = np.array([c[0] for c in count_by_occupation])
y_axis1 = np.array([c[1] for c in count_by_occupation])
```

Once we have collected the RDD of counts per occupation, we will convert it into two arrays for the *x* axis (the occupations) and the *y* axis (the counts) of our chart. The `collect` function returns the count data to us in no particular order. We need to sort the count data so that our bar chart is ordered from the lowest to the highest count.

We will achieve this by first creating two `numpy` arrays and then using the `argsort` method of `numpy` to select the elements from each array, ordered by the count data in an ascending fashion. Notice that here, we will sort both the *x* and *y* axis arrays by the *y* axis (that is, by the counts):

```
x_axis = x_axis1[np.argsort(y_axis1)]
y_axis = y_axis1[np.argsort(y_axis1)]
```

Once we have the *x* and *y* axis data for our chart, we will create the bar chart with the occupations as labels on the *x* axis and the counts as the values on the *y* axis. We will also add a few lines, such as the `plt.xticks(rotation=30)` code, to display a better-looking chart:

```
pos = np.arange(len(x_axis))
width = 1.0

ax = plt.axes()
ax.set_xticks(pos + (width / 2))
ax.set_xticklabels(x_axis)

plt.bar(pos, y_axis, width, color='lightblue')
plt.xticks(rotation=30)
fig = matplotlib.pyplot.gcf()
fig.set_size_inches(16, 10)
```

The image you have generated should look like the one here. It appears that the most prevalent occupations are **student, other, educator, administrator, engineer,** and **programmer**.

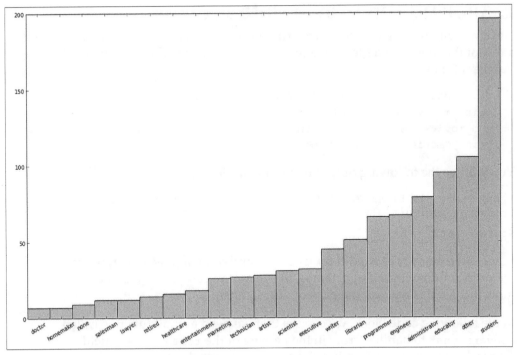

Distribution of user occupations

Spark provides a convenience method on RDDs called countByValue; this method counts the occurrences of each unique value in the RDD and returns it to the driver as a Python dict method (or a Scala or Java Map method). We can create the count_by_occupation variable using this method:

```
count_by_occupation2 = user_fields.map(lambda fields: fields[3]).
countByValue()
print "Map-reduce approach:"
print dict(count_by_occupation2)
print ""
print "countByValue approach:"
print dict(count_by_occupation)
```

You should see that the results are the same for each approach.

Exploring the movie dataset

Next, we will investigate a few properties of the movie catalogue. We can inspect a row of the movie data file, as we did for the user data earlier, and then count the number of movies:

```
movie_data = sc.textFile("/PATH/ml-100k/u.item")
print movie_data.first()
num_movies = movie_data.count()
print "Movies: %d" % num_movies
```

You will see the following output on your console:

```
1|Toy Story (1995)|01-Jan-1995||http://us.imdb.com/M/title-exact?Toy%20
Story%20(1995)|0|0|0|1|1|1|0|0|0|0|0|0|0|0|0|0|0|0
Movies: 1682
```

In the same manner as we did for user ages and occupations earlier, we can plot the distribution of movie age, that is, the year of release relative to the current date (note that for this dataset, the current year is 1998).

In the following code block, we can see that we need a small function called convert_year to handle errors in the parsing of the release date field. This is due to some bad data in one line of the movie data:

```
def convert_year(x):
  try:
    return int(x[-4:])
  except:
    return 1900 # there is a 'bad' data point with a blank year,
    which we set to 1900 and will filter out later
```

Once we have our utility function to parse the year of release, we can apply it to the movie data using a map transformation and collect the results:

```
movie_fields = movie_data.map(lambda lines: lines.split("|"))
years = movie_fields.map(lambda fields: fields[2]).map(lambda x: convert_year(x))
```

Since we have assigned the value 1900 to any error in parsing, we can filter these bad values out of the resulting data using Spark's filter transformation:

```
years_filtered = years.filter(lambda x: x != 1900)
```

This is a good example of how real-world datasets can often be messy and require a more in-depth approach to parsing data. In fact, this also illustrates why data exploration is so important, as many of these issues in data integrity and quality are picked up during this phase.

After filtering out bad data, we will transform the list of movie release years into movie ages by subtracting the current year, use countByValue to compute the counts for each movie age, and finally, plot our histogram of movie ages (again, using the hist function, where the values variable are the values of the result from countByValue, and the bins variable are the keys):

```
movie_ages = years_filtered.map(lambda yr: 1998-yr).countByValue()
values = movie_ages.values()
bins = movie_ages.keys()
hist(values, bins=bins, color='lightblue', normed=True)
fig = matplotlib.pyplot.gcf()
fig.set_size_inches(16,10)
```

You will see an image similar to the one here; it illustrates that most of the movies were released in the last few years before 1998:

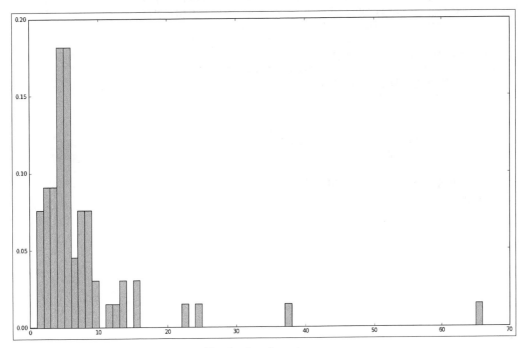

Distribution of movie ages

[63]

Exploring the rating dataset

Let's now take a look at the ratings data:

```
rating_data = sc.textFile("/PATH/ml-100k/u.data")
print rating_data.first()
num_ratings = rating_data.count()
print "Ratings: %d" % num_ratings
```

This gives us the following result:

```
196    242    3    881250949
Ratings: 100000
```

There are 100,000 ratings, and unlike the user and movie datasets, these records are split with a tab character ("\t"). As you might have guessed, we'd probably want to compute some basic summary statistics and frequency histograms for the rating values. Let's do this now:

```
rating_data = rating_data_raw.map(lambda line: line.split("\t"))
ratings = rating_data.map(lambda fields: int(fields[2]))
max_rating = ratings.reduce(lambda x, y: max(x, y))
min_rating = ratings.reduce(lambda x, y: min(x, y))
mean_rating = ratings.reduce(lambda x, y: x + y) / num_ratings
median_rating = np.median(ratings.collect())
ratings_per_user = num_ratings / num_users
ratings_per_movie = num_ratings / num_movies
print "Min rating: %d" % min_rating
print "Max rating: %d" % max_rating
print "Average rating: %2.2f" % mean_rating
print "Median rating: %d" % median_rating
print "Average # of ratings per user: %2.2f" % ratings_per_user
print "Average # of ratings per movie: %2.2f" % ratings_per_movie
```

After running these lines on your console, you will see output similar to the following result:

```
Min rating: 1
Max rating: 5
Average rating: 3.53
Median rating: 4
Average # of ratings per user: 106.00
Average # of ratings per movie: 59.00
```

We can see that the minimum rating is 1, while the maximum rating is 5. This is in line with what we expect, since the ratings are on a scale of 1 to 5.

Spark also provides a stats function for RDDs; this function contains a numeric variable (such as ratings in this case) to compute similar summary statistics:

```
ratings.stats()
```

Here is the output:

`(count: 100000, mean: 3.52986, stdev: 1.12566797076, max: 5.0, min: 1.0)`

Looking at the results, the average rating given by a user to a movie is around 3.5 and the median rating is 4, so we might expect that the distribution of ratings will be skewed towards slightly higher ratings. Let's see whether this is true by creating a bar chart of rating values using a similar procedure as we did for occupations:

```
count_by_rating = ratings.countByValue()
x_axis = np.array(count_by_rating.keys())
y_axis = np.array([float(c) for c in count_by_rating.values()])
# we normalize the y-axis here to percentages
y_axis_normed = y_axis / y_axis.sum()
pos = np.arange(len(x_axis))
width = 1.0

ax = plt.axes()
ax.set_xticks(pos + (width / 2))
ax.set_xticklabels(x_axis)

plt.bar(pos, y_axis_normed, width, color='lightblue')
plt.xticks(rotation=30)
fig = matplotlib.pyplot.gcf()
fig.set_size_inches(16, 10)
```

The preceding code should produce the following chart:

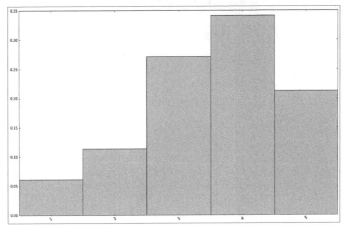

Distribution of rating values

In line with what we might have expected after seeing some summary statistics, it is clear that the distribution of ratings is skewed towards average to high ratings.

We can also look at the distribution of the number of ratings made by each user. Recall that we previously computed the `rating_data` RDD used in the preceding code by splitting the ratings with the tab character. We will now use the `rating_data` variable again in the following code.

To compute the distribution of ratings per user, we will first extract the user ID as key and rating as value from `rating_data` RDD. We will then group the ratings by user ID using Spark's `groupByKey` function:

```
user_ratings_grouped = rating_data.map(lambda fields: (int(fields[0]),
int(fields[2]))).\
    groupByKey()
```

Next, for each key (user ID), we will find the size of the set of ratings; this will give us the number of ratings for that user:

```
user_ratings_byuser = user_ratings_grouped.map(lambda (k, v): (k,
len(v)))
user_ratings_byuser.take(5)
```

We can inspect the resulting RDD by taking a few records from it; this should give us an RDD of the (user ID, number of ratings) pairs:

`[(1, 272), (2, 62), (3, 54), (4, 24), (5, 175)]`

Finally, we will plot the histogram of number of ratings per user using our favorite `hist` function:

```
user_ratings_byuser_local = user_ratings_byuser.map(lambda (k, v):
v).collect()
hist(user_ratings_byuser_local, bins=200, color='lightblue',
normed=True)
fig = matplotlib.pyplot.gcf()
fig.set_size_inches(16,10)
```

Your chart should look similar to the following screenshot. We can see that most of the users give fewer than 100 ratings. The distribution of the ratings shows, however, that there are fairly large number of users that provide hundreds of ratings.

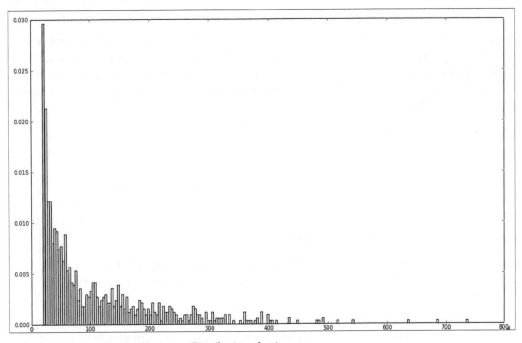

Distribution of ratings per user

We leave it to you to perform a similar analysis to create a histogram plot for the number of ratings given to each movie. Perhaps, if you're feeling adventurous, you could also extract a dataset of movie ratings by date (taken from the timestamps in the last column of the rating dataset) and chart a time series of the total number of ratings, number of unique users who gave a rating, and the number of unique movies rated, for each day.

Processing and transforming your data

Now that we have done some initial exploratory analysis of our dataset and we know a little more about the characteristics of our users and movies, what do we do next?

In order to make the raw data usable in a machine learning algorithm, we first need to clean it up and possibly transform it in various ways before extracting useful features from the transformed data. The transformation and feature extraction steps are closely linked, and in some cases, certain transformations are themselves a case of feature extraction.

We have already seen an example of the need to clean data in the movie dataset. Generally, real-world datasets contain bad data, missing data points, and outliers. Ideally, we would correct bad data; however, this is often not possible, as many datasets derive from some form of collection process that cannot be repeated (this is the case, for example, in web activity data and sensor data). Missing values and outliers are also common and can be dealt with in a manner similar to bad data. Overall, the broad options are as follows:

- **Filter out or remove records with bad or missing values**: This is sometimes unavoidable; however, this means losing the good part of a bad or missing record.

- **Fill in bad or missing data**: We can try to assign a value to bad or missing data based on the rest of the data we have available. Approaches can include assigning a zero value, assigning the global mean or median, interpolating nearby or similar data points (usually, in a time-series dataset), and so on. Deciding on the correct approach is often a tricky task and depends on the data, situation, and one's own experience.

- **Apply robust techniques to outliers**: The main issue with outliers is that they might be correct values, even though they are extreme. They might also be errors. It is often very difficult to know which case you are dealing with. Outliers can also be removed or filled in, although fortunately, there are statistical techniques (such as robust regression) to handle outliers and extreme values.

- **Apply transformations to potential outliers**: Another approach for outliers or extreme values is to apply transformations, such as a logarithmic or Gaussian kernel transformation, to features that have potential outliers, or display large ranges of potential values. These types of transformations have the effect of dampening the impact of large changes in the scale of a variable and turning a nonlinear relationship into one that is linear.

Filling in bad or missing data

We have already seen an example of filtering out bad data. Following on from the preceding code, the following code snippet applies the fill-in approach to the bad release date record by assigning a value to the data point that is equal to the median year of release:

```
years_pre_processed = movie_fields.map(lambda fields: fields[2]).
map(lambda x: convert_year(x)).collect()
years_pre_processed_array = np.array(years_pre_processed)
```

First, we will compute the mean and median year of release after selecting all the year of release data, *except* the bad data point. We will then use the numpy function, where, to find the index of the bad value in years_pre_processed_array (recall that we assigned the value 1900 to this data point). Finally, we will use this index to assign the median release year to the bad value:

```
mean_year = np.mean(years_pre_processed_array[years_pre_processed_array!=1900])
median_year = np.median(years_pre_processed_array[years_pre_processed_array!=1900])
index_bad_data = np.where(years_pre_processed_array==1900)[0][0]
years_pre_processed_array[index_bad_data] = median_year
print "Mean year of release: %d" % mean_year
print "Median year of release: %d" % median_year

print "Index of '1900' after assigning median: %s" % np.where(years_pre_processed_array == 1900)[0]
```

You should expect to see the following output:

```
Mean year of release: 1989
Median year of release: 1995
Index of '1900' after assigning median: []
```

We computed both the mean and the median year of release here. As can be seen from the output, the median release year is quite higher because of the skewed distribution of the years. While it is not always straightforward to decide on precisely which fill-in value to use for a given situation, in this case, it is certainly feasible to use the median due to this skew.

> Note that the preceding code example is, strictly speaking, not very scalable, as it requires collecting all the data to the driver. We can use Spark's mean function for numeric RDDs to compute the mean, but there is no median function available currently. We can solve this by creating our own or by computing the median on a sample of the dataset created using the `sample` function (we will see more of this in the upcoming chapters).

Extracting useful features from your data

Once we have completed the initial exploration, processing, and cleaning of our data, we are ready to get down to the business of extracting actual features from the data, with which our machine learning model can be trained.

Features refer to the variables that we use to train our model. Each row of data contains various information that we would like to extract into a training example. Almost all machine learning models ultimately work on numerical representations in the form of a **vector**; hence, we need to convert raw data into numbers.

Features broadly fall into a few categories, which are as follows:

- **Numerical features**: These features are typically real or integer numbers, for example, the user age that we used in an example earlier.
- **Categorical features**: These features refer to variables that can take one of a set of possible states at any given time. Examples from our dataset might include a user's gender or occupation or movie categories.
- **Text features**: These are features derived from the text content in the data, for example, movie titles, descriptions, or reviews.
- **Other features**: Most other types of features are ultimately represented numerically. For example, images, video, and audio can be represented as sets of numerical data. Geographical locations can be represented as latitude and longitude or geohash data.

Here we will cover numerical, categorical, and text features.

Numerical features

What is the difference between any old number and a numerical feature? Well, in reality, any numerical data can be used as an input variable. However, in a machine learning model, we learn about a vector of weights for each feature. The weights play a role in mapping feature values to an outcome or target variable (in the case of supervised learning models).

Thus, we want to use features that make sense, that is, where the model can learn the relationship between feature values and the target variable. For example, age might be a reasonable feature. Perhaps there is a direct relationship between increasing age and a certain outcome. Similarly, height is a good example of a numerical feature that can be used directly.

We will often see that numerical features are less useful in their raw form, but can be turned into representations that are more useful. Location is an example of such a case. Using raw locations (say, latitude and longitude) might not be that useful unless our data is very dense indeed, since our model might not be able to learn about a useful relationship between the raw location and an outcome. However, a relationship might exist between some aggregated or binned representation of the location (for example, a city or country) and the outcome.

Categorical features

Categorical features cannot be used as input in their raw form, as they are not numbers; instead, they are members of a set of possible values that the variable can take. In the example mentioned earlier, user occupation is a categorical variable that can take the value of student, programmer, and so on.

Such categorical variables are also known as **nominal** variables where there is no concept of order between the values of the variable. By contrast, when there is a concept of order between variables (such as the ratings mentioned earlier, where a rating of 5 is conceptually higher or better than a rating of 1), we refer to **ordinal** variables.

To transform categorical variables into a numerical representation, we can use a common approach known as **1-of-k** encoding. An approach such as 1-of-k encoding is required to represent nominal variables in a way that makes sense for machine learning tasks. Ordinal variables might be used in their raw form but are often encoded in the same way as nominal variables.

Assume that there are k possible values that the variable can take. If we assign each possible value an index from the set of 1 to k, then we can represent a given state of the variable using a binary vector of length k; here, all entries are zero, except the entry at the index that corresponds to the given state of the variable. This entry is set to one.

For example, we can collect all the possible states of the occupation variable:

```
all_occupations = user_fields.map(lambda fields: fields[3]).
distinct().collect()
all_occupations.sort()
```

We can then assign index values to each possible occupation in turn (note that we start from zero, since Python, Scala, and Java arrays all use zero-based indices):

```
idx = 0
all_occupations_dict = {}
for o in all_occupations:
    all_occupations_dict[o] = idx
    idx +=1
# try a few examples to see what "1-of-k" encoding is assigned
print "Encoding of 'doctor': %d" % all_occupations_dict['doctor']
print "Encoding of 'programmer': %d" % all_occupations_
dict['programmer']
```

You will see the following output:

Encoding of 'doctor': 2

Encoding of 'programmer': 14

Finally, we can encode the value of programmer. We will start by creating a numpy array of a length that is equal to the number of possible occupations (k in this case) and filling it with zeros. We will use the zeros function of numpy to create this array.

We will then extract the index of the word programmer and assign a value of 1 to the array value at this index:

```
K = len(all_occupations_dict)
binary_x = np.zeros(K)
k_programmer = all_occupations_dict['programmer']
binary_x[k_programmer] = 1
print "Binary feature vector: %s" % binary_x
print "Length of binary vector: %d" % K
```

This will give us the resulting binary feature vector of length 21:

```
Binary feature vector: [ 0.  0.  0.  0.  0.  0.  0.  0.  0.  0.  0.  0.
0.  0.  1.  0.  0.  0.  0.  0.  0.]
Length of binary vector: 21
```

Derived features

As we mentioned earlier, it is often useful to compute a derived feature from one or more available variables. We hope that the derived feature can add more information than only using the variable in its raw form.

For instance, we can compute the average rating given by each user to all the movies they rated. This would be a feature that could provide a *user-specific* intercept in our model (in fact, this is a commonly used approach in recommendation models). We have taken the raw rating data and created a new feature that can allow us to learn a better model.

Examples of features derived from raw data include computing average values, median values, variances, sums, differences, maximums or minimums, and counts. We have already seen a case of this when we created a new `movie age` feature from the year of release of the movie and the current year. Often, the idea behind using these transformations is to summarize the numerical data in some way that might make it easier for a model to learn.

It is also common to transform numerical features into categorical features, for example, by binning features. Common examples of this include variables such as age, geolocation, and time.

Transforming timestamps into categorical features

To illustrate how to derive categorical features from numerical data, we will use the times of the ratings given by users to movies. These are in the form of Unix timestamps. We can use Python's `datetime` module to extract the date and time from the timestamp and, in turn, extract the `hour` of the day. This will result in an RDD of the hour of the day for each rating.

We will need a function to extract a `datetime` representation of the rating timestamp (in seconds); we will create this function now:

```
def extract_datetime(ts):
    import datetime
    return datetime.datetime.fromtimestamp(ts)
```

We will again use the `rating_data` RDD that we computed in the earlier examples as our starting point.

First, we will use a `map` transformation to extract the timestamp field, converting it to a Python `int` datatype. We will then apply our `extract_datetime` function to each timestamp and extract the hour from the resulting `datetime` object:

```
timestamps = rating_data.map(lambda fields: int(fields[3]))
hour_of_day = timestamps.map(lambda ts: extract_datetime(ts).hour)
hour_of_day.take(5)
```

If we take the first five records of the resulting RDD, we will see the following output:

`[17, 21, 9, 7, 7]`

We have transformed the raw time data into a categorical feature that represents the hour of the day in which the rating was given.

Now, say that we decide this is too coarse a representation. Perhaps we want to further refine the transformation. We can assign each hour-of-the-day value into a defined bucket that represents a time of day.

For example, we can say that morning is from 7 a.m. to 11 a.m., while lunch is from 11 a.m. to 1 a.m., and so on. Using these buckets, we can create a function to assign a time of day, given the hour of the day as input:

```
def assign_tod(hr):
  times_of_day = {
    'morning' : range(7, 12),
    'lunch' : range(12, 14),
    'afternoon' : range(14, 18),
    'evening' : range(18, 23),
    'night' : range(23, 7)
  }
  for k, v in times_of_day.iteritems():
    if hr in v:
      return k
```

Now, we will apply the `assign_tod` function to the hour of each rating event contained in the `hour_of_day` RDD:

```
time_of_day = hour_of_day.map(lambda hr: assign_tod(hr))
time_of_day.take(5)
```

If we again take the first five records of this new RDD, we will see the following transformed values:

`['afternoon', 'evening', 'morning', 'morning', 'morning']`

We have now transformed the timestamp variable (which can take on thousands of values and is probably not useful to a model in its raw form) into hours (taking on 24 values) and then into a time of day (taking on five possible values). Now that we have a categorical feature, we can use the same 1-of-k encoding method outlined earlier to generate a binary feature vector.

Text features

In some ways, text features are a form of categorical and derived features. Let's take the example of the description for a movie (which we do not have in our dataset). Here, the raw text could not be used directly, even as a categorical feature, since there are virtually unlimited possible combinations of words that could occur if each piece of text was a possible value. Our model would almost never see two occurrences of the same feature and would not be able to learn effectively. Therefore, we would like to turn raw text into a form that is more amenable to machine learning.

There are numerous ways of dealing with text, and the field of natural language processing is dedicated to processing, representing, and modeling textual content. A full treatment is beyond the scope of this book, but we will introduce a simple and standard approach for text-feature extraction; this approach is known as the **bag-of-words** representation.

The bag-of-words approach treats a piece of text content as a set of the words, and possibly numbers, in the text (these are often referred to as terms). The process of the bag-of-words approach is as follows:

- **Tokenization**: First, some form of tokenization is applied to the text to split it into a set of tokens (generally words, numbers, and so on). An example of this is simple whitespace tokenization, which splits the text on each space and might remove punctuation and other characters that are not alphabetical or numerical.
- **Stop word removal**: Next, it is usual to remove very common words such as "the", "and", and "but" (these are known as **stop words**).
- **Stemming**: The next step can include stemming, which refers to taking a term and reducing it to its base form or stem. A common example is plural terms becoming singular (for example, dogs becomes dog and so on). There are many approaches to stemming, and text-processing libraries often contain various stemming algorithms.

- **Vectorization**: The final step is turning the processed terms into a vector representation. The simplest form is, perhaps, a binary vector representation, where we assign a value of one if a term exists in the text and zero if it does not. This is essentially identical to the categorical 1-of-k encoding we encountered earlier. Like 1-of-k encoding, this requires a dictionary of terms mapping a given term to an index number. As you might gather, there are potentially millions of individual possible terms (even after stop word removal and stemming). Hence, it becomes critical to use a sparse vector representation where only the fact that a term is present is stored, to save memory and disk space as well as compute time.

In *Chapter 9, Advanced Text Processing with Spark*, we will cover more complex text processing and feature extraction, including methods to weight terms; these methods go beyond the basic binary encoding we saw earlier.

Simple text feature extraction

To show an example of extracting textual features in the binary vector representation, we can use the movie titles that we have available.

First, we will create a function to strip away the year of release for each movie, if the year is present, leaving only the title of the movie.

We will use Python's regular expression module, `re`, to search for the year between parentheses in the movie titles. If we find a match with this regular expression, we will extract only the title up to the index of the first match (that is, the index in the title string of the opening parenthesis). This is done with the following `raw[:grps.start()]` code snippet:

```
def extract_title(raw):
  import re
  # this regular expression finds the non-word (numbers) between
  parentheses
  grps = re.search("\((\w+)\)", raw)
  if grps:
    # we take only the title part, and strip the trailing
    whitespace from the remaining text, below
    return raw[:grps.start()].strip()
  else:
    return raw
```

Next, we will extract the raw movie titles from the `movie_fields` RDD:

```
raw_titles = movie_fields.map(lambda fields: fields[1])
```

We can test out our `extract_title` function on the first five raw titles as follows:

```
for raw_title in raw_titles.take(5):
    print extract_title(raw_title)
```

We can verify that our function works by inspecting the results, which should look like this:

```
Toy Story
GoldenEye
Four Rooms
Get Shorty
Copycat
```

We would then like to apply our function to the raw titles and apply a tokenization scheme to the extracted titles to convert them to terms. We will use the simple whitespace tokenization we covered earlier:

```
movie_titles = raw_titles.map(lambda m: extract_title(m))
# next we tokenize the titles into terms. We'll use simple whitespace
tokenization
title_terms = movie_titles.map(lambda t: t.split(" "))
print title_terms.take(5)
```

Applying this simple tokenization gives the following result:

```
[[u'Toy', u'Story'], [u'GoldenEye'], [u'Four', u'Rooms'], [u'Get', u'Shorty'], [u'Copycat']]
```

We can see that we have split each title on spaces so that each word becomes a token.

> Here, we do not cover details such as converting text to lowercase, removing non-word or non-numerical characters such as punctuation and special characters, removing stop words, and stemming. These steps will be important in a real-world application. We will cover many of these topics in *Chapter 9, Advanced Text Processing with Spark*.
>
> This additional processing can be done fairly simply using string functions, regular expressions, and the Spark API (apart from stemming). Perhaps you would like to give it a try!

In order to assign each term to an index in our vector, we need to create the term dictionary, which maps each term to an integer index.

First, we will use Spark's `flatMap` function (highlighted in the following code snippet) to expand the list of strings in each record of the `title_terms` RDD into a new RDD of strings where each record is a term called `all_terms`.

We can then collect all the unique terms and assign indexes in exactly the same way that we did for the 1-of-k encoding of user occupations earlier:

```
# next we would like to collect all the possible terms, in order to
build out dictionary of term <-> index mappings
all_terms = title_terms.flatMap(lambda x: x).distinct().collect()
# create a new dictionary to hold the terms, and assign the "1-of-k"
indexes
idx = 0
all_terms_dict = {}
for term in all_terms:
   all_terms_dict[term] = idx
   idx +=1
```

We can print out the total number of unique terms and test out our term mapping on a few terms:

```
print "Total number of terms: %d" % len(all_terms_dict)
print "Index of term 'Dead': %d" % all_terms_dict['Dead']
print "Index of term 'Rooms': %d" % all_terms_dict['Rooms']
```

This will result in the following output:

```
Total number of terms: 2645
Index of term 'Dead': 147
Index of term 'Rooms': 1963
```

We can also achieve the same result more efficiently using Spark's `zipWithIndex` function. This function takes an RDD of values and merges them together with an index to create a new RDD of key-value pairs, where the key will be the term and the value will be the index in the term dictionary. We will use `collectAsMap` to collect the key-value RDD to the driver as a Python `dict` method:

```
all_terms_dict2 = title_terms.flatMap(lambda x: x).distinct().
zipWithIndex().collectAsMap()
print "Index of term 'Dead': %d" % all_terms_dict2['Dead']
print "Index of term 'Rooms': %d" % all_terms_dict2['Rooms']
```

The output is as follows:

```
Index of term 'Dead': 147
Index of term 'Rooms': 1963
```

The final step is to create a function that converts a set of terms into a sparse vector representation. To do this, we will create an empty sparse matrix with one row and a number of columns equal to the total number of terms in our dictionary. We will then step through each term in the input list of terms and check whether this term is in our term dictionary. If it is, we assign a value of 1 to the vector at the index that corresponds to the term in our dictionary mapping:

```
# this function takes a list of terms and encodes it as a scipy sparse
vector using an approach
# similar to the 1-of-k encoding
def create_vector(terms, term_dict):
   from scipy import sparse as sp
     num_terms = len(term_dict)
     x = sp.csc_matrix((1, num_terms))
     for t in terms:
        if t in term_dict:
           idx = term_dict[t]
           x[0, idx] = 1
     return x
```

Once we have our function, we will apply it to each record in our RDD of extracted terms:

```
all_terms_bcast = sc.broadcast(all_terms_dict)
term_vectors = title_terms.map(lambda terms: create_vector(terms, all_
terms_bcast.value))
term_vectors.take(5)
```

We can then inspect the first few records of our new RDD of sparse vectors:

```
[<1x2645 sparse matrix of type '<type 'numpy.float64'>'
  with 2 stored elements in Compressed Sparse Column format>,
 <1x2645 sparse matrix of type '<type 'numpy.float64'>'
  with 1 stored elements in Compressed Sparse Column format>,
 <1x2645 sparse matrix of type '<type 'numpy.float64'>'
  with 2 stored elements in Compressed Sparse Column format>,
 <1x2645 sparse matrix of type '<type 'numpy.float64'>'
  with 2 stored elements in Compressed Sparse Column format>,
 <1x2645 sparse matrix of type '<type 'numpy.float64'>'
  with 1 stored elements in Compressed Sparse Column format>]
```

We can see that each movie title has now been transformed into a sparse vector. We can see that the titles where we extracted two terms have two non-zero entries in the vector, titles where we extracted only one term have one non-zero entry, and so on.

> Note the use of Spark's `broadcast` method in the preceding example code to create a broadcast variable that contains the term dictionary. In real-world applications, such term dictionaries can be extremely large, so using a broadcast variable is advisable.

Normalizing features

Once the features have been extracted into the form of a vector, a common preprocessing step is to normalize the numerical data. The idea behind this is to transform each numerical feature in a way that scales it to a standard size. We can perform different kinds of normalization, which are as follows:

- **Normalize a feature**: This is usually a transformation applied to an individual feature across the dataset, for example, subtracting the mean (*centering* the feature) or applying the standard normal transformation (such that the feature has a mean of zero and a standard deviation of 1).
- **Normalize a feature vector**: This is usually a transformation applied to all features in a given row of the dataset such that the resulting feature vector has a normalized length. That is, we will ensure that each feature in the vector is scaled such that the vector has a norm of 1 (typically, on an L1 or L2 norm).

We will use the second case as an example. We can use the `norm` function of `numpy` to achieve the vector normalization by first computing the L2 norm of a random vector and then dividing each element in the vector by this norm to create our normalized vector:

```
np.random.seed(42)
x = np.random.randn(10)
norm_x_2 = np.linalg.norm(x)
normalized_x = x / norm_x_2
print "x:\n%s" % x
print "2-Norm of x: %2.4f" % norm_x_2
print "Normalized x:\n%s" % normalized_x
print "2-Norm of normalized_x: %2.4f" %
np.linalg.norm(normalized_x)
```

This should give the following result (note that in the preceding code snippet, we set the random seed equal to 42 so that the result will always be the same):

```
x: [ 0.49671415 -0.1382643  0.64768854  1.52302986 -0.23415337 -0.23413696
 1.57921282  0.76743473 -0.46947439  0.54256004]
2-Norm of x: 2.5908
Normalized x: [ 0.19172213 -0.05336737  0.24999534  0.58786029 -0.09037871
 -0.09037237  0.60954584  0.29621508 -0.1812081   0.20941776]
2-Norm of normalized_x: 1.0000
```

Using MLlib for feature normalization

Spark provides some built-in functions for feature scaling and standardization in its MLlib machine learning library. These include `StandardScaler`, which applies the standard normal transformation, and `Normalizer`, which applies the same feature vector normalization we showed you in our preceding example code.

We will explore the use of these methods in the upcoming chapters, but for now, let's simply compare the results of using MLlib's `Normalizer` to our own results:

```
from pyspark.mllib.feature import Normalizer
normalizer = Normalizer()
vector = sc.parallelize([x])
```

After importing the required class, we will instantiate `Normalizer` (by default, it will use the L2 norm as we did earlier). Note that as in most situations in Spark, we need to provide `Normalizer` with an RDD as input (it contains numpy arrays or MLlib vectors); hence, we will create a single-element RDD from our vector x for illustrative purposes.

We will then use the `transform` function of `Normalizer` on our RDD. Since the RDD only has one vector in it, we will return our vector to the driver by calling `first` and finally by calling the `toArray` function to convert the vector back into a numpy array:

```
normalized_x_mllib = normalizer.transform(vector).first().toArray()
```

Finally, we can print out the same details as we did previously, comparing the results:

```
print "x:\n%s" % x
print "2-Norm of x: %2.4f" % norm_x_2
print "Normalized x MLlib:\n%s" % normalized_x_mllib
print "2-Norm of normalized_x_mllib: %2.4f" % np.linalg.norm(normalized_x_mllib)
```

You will end up with exactly the same normalized vector as we did with our own code. However, using MLlib's built-in methods is certainly more convenient and efficient than writing our own functions!

Using packages for feature extraction

While we have covered many different approaches to feature extraction, it will be rather painful to have to create the code to perform these common tasks each and every time. Certainly, we can create our own reusable code libraries for this purpose; however, fortunately, we can rely on the existing tools and packages.

Since Spark supports Scala, Java, and Python bindings, we can use packages available in these languages that provide sophisticated tools to process and extract features and represent them as vectors. A few examples of packages for feature extraction include scikit-learn, gensim, scikit-image, matplotlib, and NLTK in Python; OpenNLP in Java; and Breeze and Chalk in Scala. In fact, Breeze has been part of Spark MLlib since version 1.0, and we will see how to use some Breeze functionality for linear algebra in the later chapters.

Summary

In this chapter, we saw how to find common, publicly-available datasets that can be used to test various machine learning models. You learned how to load, process, and clean data, as well as how to apply common techniques to transform raw data into feature vectors that can be used as training examples for our models.

In the next chapter, you will learn the basics of recommender systems and explore how to create a recommendation model, use the model to make predictions, and evaluate the model.

Building a Recommendation Engine with Spark

Now that you have learned the basics of data processing and feature extraction, we will move on to explore individual machine learning models in detail, starting with recommendation engines.

Recommendation engines are probably among the best types of machine learning model known to the general public. Even if people do not know exactly what a recommendation engine is, they have most likely experienced one through the use of popular websites such as Amazon, Netflix, YouTube, Twitter, LinkedIn, and Facebook. Recommendations are a core part of all these businesses, and in some cases, they drive significant percentages of their revenue.

The idea behind recommendation engines is to predict what people might like and to uncover relationships between items to aid in the discovery process (in this way, it is similar and, in fact, often complementary to search engines, which also play a role in discovery). However, unlike search engines, recommendation engines try to present people with relevant content that they did not necessarily search for or that they might not even have heard of.

Typically, a recommendation engine tries to model the connections between users and some type of item. In our MovieStream scenario from *Chapter 2, Designing a Machine Learning System*, for example, we could use a recommendation engine to show our users movies that they might enjoy. If we can do this well, we could keep our users engaged using our service, which is good for both our users and us. Similarly, if we can do a good job of showing our users movies related to a given movie, we could aid in discovery and navigation on our site, again improving our users' experience, engagement, and the relevance of our content to them.

However, recommendation engines are not limited to movies, books, or products. The techniques we will explore in this chapter can be applied to just about any user-to-item relationship as well as user-to-user connections, such as those found on social networks, allowing us to make recommendations such as people you may know or who to follow.

Recommendation engines are most effective in two general scenarios (which are not mutually exclusive). They are explained here:

- **Large number of available options for users**: When there are a very large number of available items, it becomes increasingly difficult for the user to find something they want. Searching can help when the user knows what they are looking for, but often, the right item might be something previously unknown to them. In this case, being recommended relevant items, that the user may not already know about, can help them discover new items.
- **A significant degree of personal taste involved**: When personal taste plays a large role in selection, recommendation models, which often utilize a wisdom of the crowd approach, can be helpful in discovering items based on the behavior of others that have similar taste profiles.

In this chapter, we will:

- Introduce the various types of recommendation engines
- Build a recommendation model using data about user preferences
- Use the trained model to compute recommendations for a given user as well compute similar items for a given item (that is, related items)
- Apply standard evaluation metrics to the model that we created to measure how well it performs in terms of predictive capability

Types of recommendation models

Recommender systems are widely studied, and there are many approaches used, but there are two that are probably most prevalent: content-based filtering and collaborative filtering. Recently, other approaches such as ranking models have also gained in popularity. In practice, many approaches are hybrids, incorporating elements of many different methods into a model or combination of models.

Content-based filtering

Content-based methods try to use the content or attributes of an item, together with some notion of similarity between two pieces of content, to generate items similar to a given item. These attributes are often textual content (such as titles, names, tags, and other metadata attached to an item), or in the case of media, they could include other features of the item, such as attributes extracted from audio and video content.

In a similar manner, user recommendations can be generated based on attributes of users or user profiles, which are then matched to item attributes using the same measure of similarity. For example, a user can be represented by the combined attributes of the items they have interacted with. This becomes their user profile, which is then compared to item attributes to find items that match the user profile.

Collaborative filtering

Collaborative filtering is a form of wisdom of the crowd approach where the set of preferences of many users with respect to items is used to generate estimated preferences of users for items with which they have not yet interacted. The idea behind this is the notion of similarity.

In a user-based approach, if two users have exhibited similar preferences (that is, patterns of interacting with the same items in broadly the same way), then we would assume that they are similar to each other in terms of taste. To generate recommendations for unknown items for a given user, we can use the known preferences of other users that exhibit similar behavior. We can do this by selecting a set of similar users and computing some form of combined score based on the items they have shown a preference for. The overall logic is that if others have tastes similar to a set of items, these items would tend to be good candidates for recommendation.

We can also take an item-based approach that computes some measure of similarity between items. This is usually based on the existing user-item preferences or ratings. Items that tend to be rated the same by similar users will be classed as similar under this approach. Once we have these similarities, we can represent a user in terms of the items they have interacted with and find items that are similar to these known items, which we can then recommend to the user. Again, a set of items similar to the known items is used to generate a combined score to estimate for an unknown item.

The user- and item-based approaches are usually referred to as nearest-neighbor models, since the estimated scores are computed based on the set of most similar users or items (that is, their neighbors).

Finally, there are many model-based methods that attempt to model the user-item preferences themselves so that new preferences can be estimated directly by applying the model to unknown user-item combinations.

Matrix factorization

Since Spark's recommendation models currently only include an implementation of matrix factorization, we will focus our attention on this class of models. This focus is with good reason; however, these types of models have consistently been shown to perform extremely well in collaborative filtering and were among the best models in well-known competitions such as the Netflix prize.

> For more information on and a brief overview of the performance of the best algorithms for the Netflix prize, see http://techblog.netflix.com/2012/04/netflix-recommendations-beyond-5-stars.html.

Explicit matrix factorization

When we deal with data that consists of preferences of users that are provided by the users themselves, we refer to explicit preference data. This includes, for example, ratings, thumbs up, likes, and so on that are given by users to items.

We can take these ratings and form a two-dimensional matrix with users as rows and items as columns. Each entry represents a rating given by a user to a certain item. Since in most cases, each user has only interacted with a relatively small set of items, this matrix has only a few non-zero entries (that is, it is very sparse).

As a simple example, let's assume that we have the following user ratings for a set of movies:

```
Tom, Star Wars, 5
Jane, Titanic, 4
Bill, Batman, 3
Jane, Star Wars, 2
Bill, Titanic, 3
```

We will form the following ratings matrix:

User / Item	Batman	Star Wars	Titanic
Bill	3	3	
Jane		2	4
Tom		5	

A simple movie-rating matrix

Matrix factorization (or matrix completion) attempts to directly model this user-item matrix by representing it as a product of two smaller matrices of lower dimension. Thus, it is a dimensionality-reduction technique. If we have **U** users and **I** items, then our user-item matrix is of dimension U x I and might look something like the one shown in the following diagram:

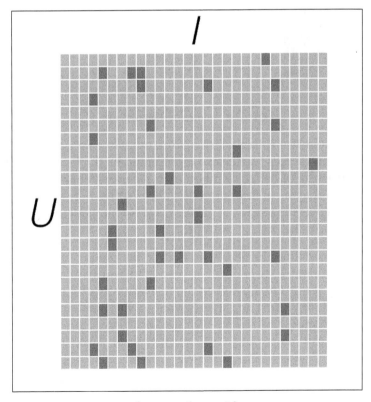

A sparse ratings matrix

If we want to find a lower dimension (low-rank) approximation to our user-item matrix with the dimension **k**, we would end up with two matrices: one for users of size U x k and one for items of size I x k. These are known as factor matrices. If we multiply these two factor matrices, we would reconstruct an approximate version of the original ratings matrix. Note that while the original ratings matrix is typically very sparse, each factor matrix is dense, as shown in the following diagram:

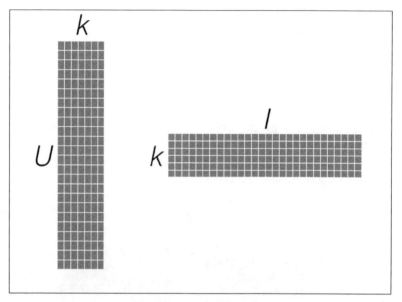

The user- and item-factor matrices

These models are often also called latent feature models, as we are trying to discover some form of hidden features (which are represented by the factor matrices) that account for the structure of behavior inherent in the user-item rating matrix. While the latent features or factors are not directly interpretable, they might, perhaps, represent things such as the tendency of a user to like movies from a certain director, genre, style, or group of actors, for example.

As we are directly modeling the user-item matrix, the prediction in these models is relatively straightforward: to compute a predicted rating for a user and item, we compute the vector dot product between the relevant row of the user-factor matrix (that is, the user's factor vector) and the relevant row of the item-factor matrix (that is, the item's factor vector).

This is illustrated with the highlighted vectors in the following diagram:

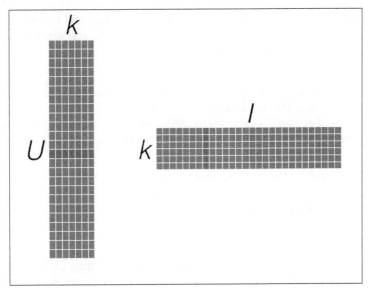

Computing recommendations from user- and item-factor vectors

To find out the similarity between two items, we can use the same measures of similarity as we would use in the nearest-neighbor models, except that we can use the factor vectors directly by computing the similarity between two item-factor vectors, as illustrated in the following diagram:

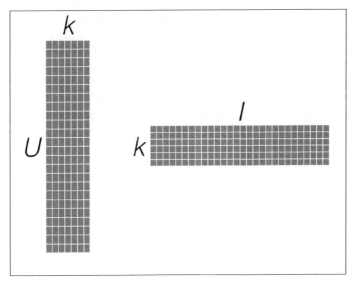

Computing similarity with item-factor vectors

The benefit of factorization models is the relative ease of computing recommendations once the model is created. However, for very large user and itemsets, this can become a challenge as it requires storage and computation across potentially many millions of user- and item-factor vectors. Another advantage, as mentioned earlier, is that they tend to offer very good performance.

> Projects such as Oryx (https://github.com/OryxProject/oryx) and Prediction.io (https://github.com/PredictionIO/PredictionIO) focus on model serving for large-scale models, including recommenders based on matrix factorization.

On the down side, factorization models are relatively more complex to understand and interpret compared to nearest-neighbor models and are often more computationally intensive during the model's training phase.

Implicit matrix factorization

So far, we have dealt with explicit preferences such as ratings. However, much of the preference data that we might be able to collect is implicit feedback, where the preferences between a user and item are not given to us, but are, instead, implied from the interactions they might have with an item. Examples include binary data (such as whether a user viewed a movie, whether they purchased a product, and so on) as well as count data (such as the number of times a user watched a movie).

There are many different approaches to deal with implicit data. MLlib implements a particular approach that treats the input rating matrix as two matrices: a binary preference matrix, **P**, and a matrix of confidence weights, **C**.

For example, let's assume that the user-movie ratings we saw previously were, in fact, the number of times each user had viewed that movie. The two matrices would look something like ones shown in the following screenshot. Here, the matrix **P** informs us that a movie was viewed by a user, and the matrix **C** represents the confidence weighting, in the form of the view counts — generally, the more a user has watched a movie, the higher the confidence that they actually like it.

P					C			
User / Item	Batman	Star Wars	Titanic		User / Item	Batman	Star Wars	Titanic
Bill	1	1			Bill	3	3	
Jane		1	1		Jane		2	4
Tom		1			Tom		5	

Representation of an implicit preference and confidence matrix

The implicit model still creates a user- and item-factor matrix. In this case, however, the matrix that the model is attempting to approximate is not the overall ratings matrix but the preference matrix P. If we compute a recommendation by calculating the dot product of a user- and item-factor vector, the score will not be an estimate of a rating directly. It will rather be an estimate of the preference of a user for an item (though not strictly between 0 and 1, these scores will generally be fairly close to a scale of 0 to 1).

Alternating least squares

Alternating Least Squares (**ALS**) is an optimization technique to solve matrix factorization problems; this technique is powerful, achieves good performance, and has proven to be relatively easy to implement in a parallel fashion. Hence, it is well suited for platforms such as Spark. At the time of writing this book, it is the only recommendation model implemented in MLlib.

ALS works by iteratively solving a series of least squares regression problems. In each iteration, one of the user- or item-factor matrices is treated as fixed, while the other one is updated using the fixed factor and the rating data. Then, the factor matrix that was solved for is, in turn, treated as fixed, while the other one is updated. This process continues until the model has converged (or for a fixed number of iterations).

 Spark's documentation for collaborative filtering contains references to the papers that underlie the ALS algorithms implemented each component of explicit and implicit data. You can view the documentation at http://spark.apache.org/docs/latest/mllib-collaborative-filtering.html.

Extracting the right features from your data

In this section, we will use explicit rating data, without additional user or item metadata or other information related to the user-item interactions. Hence, the features that we need as inputs are simply the user IDs, movie IDs, and the ratings assigned to each user and movie pair.

Extracting features from the MovieLens 100k dataset

Start the Spark shell in the Spark base directory, ensuring that you provide enough memory via the -driver-memory option:

```
>./bin/spark-shell –driver-memory 4g
```

In this example, we will use the same MovieLens dataset that we used in the previous chapter. Use the directory in which you placed the MovieLens 100k dataset as the input path in the following code.

First, let's inspect the raw ratings dataset:

```
val rawData = sc.textFile("/PATH/ml-100k/u.data")
rawData.first()
```

You will see output similar to these lines of code:

```
14/03/30 11:42:41 WARN NativeCodeLoader: Unable to load native-hadoop library for your platform... using builtin-java classes where applicable
14/03/30 11:42:41 WARN LoadSnappy: Snappy native library not loaded
14/03/30 11:42:41 INFO FileInputFormat: Total input paths to process : 1
14/03/30 11:42:41 INFO SparkContext: Starting job: first at <console>:15
14/03/30 11:42:41 INFO DAGScheduler: Got job 0 (first at <console>:15) with 1 output partitions (allowLocal=true)
14/03/30 11:42:41 INFO DAGScheduler: Final stage: Stage 0 (first at <console>:15)
14/03/30 11:42:41 INFO DAGScheduler: Parents of final stage: List()
14/03/30 11:42:41 INFO DAGScheduler: Missing parents: List()
14/03/30 11:42:41 INFO DAGScheduler: Computing the requested partition locally
```

```
14/03/30 11:42:41 INFO HadoopRDD: Input split: file:/Users/Nick/
workspace/datasets/ml-100k/u.data:0+1979173
14/03/30 11:42:41 INFO SparkContext: Job finished: first at <console>:15,
took 0.030533 s
res0: String = 196    242    3    881250949
```

Recall that this dataset consisted of the `user id`, `movie id`, `rating`, `timestamp` fields separated by a tab (`"\t"`) character. We don't need the time when the rating was made to train our model, so let's simply extract the first three fields:

```
val rawRatings = rawData.map(_.split("\t").take(3))
```

We will first split each record on the `"\t"` character, which gives us an `Array[String]` array. We will then use Scala's `take` function to keep only the first 3 elements of the array, which correspond to `user id`, `movie id`, and `rating`, respectively.

We can inspect the first record of our new RDD by calling `rawRatings.first()`, which collects just the first record of the RDD back to the driver program. This will result in the following output:

```
14/03/30 12:24:00 INFO SparkContext: Starting job: first at <console>:21
14/03/30 12:24:00 INFO DAGScheduler: Got job 1 (first at <console>:21)
with 1 output partitions (allowLocal=true)
14/03/30 12:24:00 INFO DAGScheduler: Final stage: Stage 1 (first at
<console>:21)
14/03/30 12:24:00 INFO DAGScheduler: Parents of final stage: List()
14/03/30 12:24:00 INFO DAGScheduler: Missing parents: List()
14/03/30 12:24:00 INFO DAGScheduler: Computing the requested partition
locally
14/03/30 12:24:00 INFO HadoopRDD: Input split: file:/Users/Nick/
workspace/datasets/ml-100k/u.data:0+1979173
14/03/30 12:24:00 INFO SparkContext: Job finished: first at <console>:21,
took 0.00391 s
res6: Array[String] = Array(196, 242, 3)
```

We will use Spark's MLlib library to train our model. Let's take a look at what methods are available for us to use and what input is required. First, import the ALS model from MLlib:

```
import org.apache.spark.mllib.recommendation.ALS
```

On the console, we can inspect the available methods on the ALS object using tab completion. Type in `ALS.` (note the dot) and then press the *Tab* key. You should see the autocompletion of the methods:

```
ALS.
asInstanceOf     isInstanceOf     main          toString         train
trainImplicit
```

The method we want to use is `train`. If we type `ALS.train` and hit *Enter*, we will get an error. However, this error will tell us what the method signature looks like:

```
ALS.train
<console>:12: error: ambiguous reference to overloaded definition,
both method train in object ALS of type (ratings: org.apache.spark.rdd.
RDD[org.apache.spark.mllib.recommendation.Rating], rank: Int, iterations:
Int)org.apache.spark.mllib.recommendation.MatrixFactorizationModel
and  method train in object ALS of type (ratings: org.apache.spark.
rdd.RDD[org.apache.spark.mllib.recommendation.Rating], rank: Int,
iterations: Int, lambda: Double)org.apache.spark.mllib.recommendation.
MatrixFactorizationModel
match expected type ?
              ALS.train
                  ^
```

So, we can see that at a minimum, we need to provide the input arguments, `ratings`, `rank`, and `iterations`. The second method also requires an argument called `lambda`. We'll cover these three shortly, but let's take a look at the `ratings` argument. First, let's import the `Rating` class that it references and use a similar approach to find out what an instance of `Rating` requires, by typing in `Rating()` and hitting *Enter*:

```
import org.apache.spark.mllib.recommendation.Rating
Rating()
<console>:13: error: not enough arguments for method apply: (user: Int,
product: Int, rating: Double)org.apache.spark.mllib.recommendation.Rating
in object Rating.
Unspecified value parameters user, product, rating.
              Rating()
                    ^
```

As we can see from the preceding output, we need to provide the ALS model with an RDD that consists of `Rating` records. A `Rating` class, in turn, is just a wrapper around `user id`, `movie id` (called `product` here), and the actual `rating` arguments. We'll create our rating dataset using the `map` method and transforming the array of IDs and ratings into a `Rating` object:

```
val ratings = rawRatings.map { case Array(user, movie, rating) =>
Rating(user.toInt, movie.toInt, rating.toDouble) }
```

> Notice that we need to use `toInt` or `toDouble` to convert the raw rating data (which was extracted as `Strings` from the text file) to `Int` or `Double` numeric inputs. Also, note the use of a `case` statement that allows us to extract the relevant variable names and use them directly (this saves us from having to use something like `val user = ratings(0)`).
>
> For more on Scala case statements and pattern matching as used here, take a look at http://docs.scala-lang.org/tutorials/tour/pattern-matching.html.

We now have an `RDD[Rating]` that we can verify by calling:

```
ratings.first()
14/03/30 12:32:48 INFO SparkContext: Starting job: first at <console>:24
14/03/30 12:32:48 INFO DAGScheduler: Got job 2 (first at <console>:24) with 1 output partitions (allowLocal=true)
14/03/30 12:32:48 INFO DAGScheduler: Final stage: Stage 2 (first at <console>:24)
14/03/30 12:32:48 INFO DAGScheduler: Parents of final stage: List()
14/03/30 12:32:48 INFO DAGScheduler: Missing parents: List()
14/03/30 12:32:48 INFO DAGScheduler: Computing the requested partition locally
14/03/30 12:32:48 INFO HadoopRDD: Input split: file:/Users/Nick/workspace/datasets/ml-100k/u.data:0+1979173
14/03/30 12:32:48 INFO SparkContext: Job finished: first at <console>:24, took 0.003752 s
res8: org.apache.spark.mllib.recommendation.Rating = Rating(196,242,3.0)
```

Training the recommendation model

Once we have extracted these simple features from our raw data, we are ready to proceed with model training; MLlib takes care of this for us. All we have to do is provide the correctly-parsed input RDD we just created as well as our chosen model parameters.

Training a model on the MovieLens 100k dataset

We're now ready to train our model! The other inputs required for our model are as follows:

- `rank`: This refers to the number of factors in our ALS model, that is, the number of hidden features in our low-rank approximation matrices. Generally, the greater the number of factors, the better, but this has a direct impact on memory usage, both for computation and to store models for serving, particularly for large number of users or items. Hence, this is often a trade-off in real-world use cases. A rank in the range of 10 to 200 is usually reasonable.

- `iterations`: This refers to the number of iterations to run. While each iteration in ALS is guaranteed to decrease the reconstruction error of the ratings matrix, ALS models will converge to a reasonably good solution after relatively few iterations. So, we don't need to run for too many iterations in most cases (around 10 is often a good default).

- `lambda`: This parameter controls the regularization of our model. Thus, `lambda` controls over fitting. The higher the value of `lambda`, the more is the regularization applied. What constitutes a sensible value is very dependent on the size, nature, and sparsity of the underlying data, and as with almost all machine learning models, the regularization parameter is something that should be tuned using out-of-sample test data and cross-validation approaches.

We'll use `rank` of 50, 10 iterations, and a lambda parameter of 0.01 to illustrate how to train our model:

```
val model = ALS.train(ratings, 50, 10, 0.01)
```

Chapter 4

This returns a `MatrixFactorizationModel` object, which contains the user and item factors in the form of an RDD of `(id, factor)` pairs. These are called `userFeatures` and `productFeatures`, respectively. For example:

```
model.userFeatures
```

You will see the output as:

```
res14: org.apache.spark.rdd.RDD[(Int, Array[Double])] = FlatMappedRDD[659] at flatMap at ALS.scala:231
```

We can see that the factors are in the form of an `Array[Double]`.

Note that the operations used in MLlib's ALS implementation are lazy transformations, so the actual computation will only be performed once we call some sort of action on the resulting RDDs of the user and item factors. We can force the computation using a Spark action such as `count`:

```
model.userFeatures.count
```

This will trigger the computation, and we will see a quite a bit of output text similar to the following lines of code:

```
14/03/30 13:10:40 INFO SparkContext: Starting job: count at <console>:26
14/03/30 13:10:40 INFO DAGScheduler: Registering RDD 665 (map at ALS.scala:147)
14/03/30 13:10:40 INFO DAGScheduler: Registering RDD 664 (map at ALS.scala:146)
14/03/30 13:10:40 INFO DAGScheduler: Registering RDD 674 (mapPartitionsWithIndex at ALS.scala:164)
...
14/03/30 13:10:45 INFO SparkContext: Job finished: count at <console>:26, took 5.068255 s
res16: Long = 943
```

If we call `count` for the movie factors, we will see the following output:

```
model.productFeatures.count
14/03/30 13:15:21 INFO SparkContext: Starting job: count at <console>:26
14/03/30 13:15:21 INFO DAGScheduler: Got job 10 (count at <console>:26) with 1 output partitions (allowLocal=false)
14/03/30 13:15:21 INFO DAGScheduler: Final stage: Stage 165 (count at <console>:26)
14/03/30 13:15:21 INFO DAGScheduler: Parents of final stage: List(Stage 169, Stage 166)
```

```
14/03/30 13:15:21 INFO DAGScheduler: Missing parents: List()
14/03/30 13:15:21 INFO DAGScheduler: Submitting Stage 165
(FlatMappedRDD[883] at flatMap at ALS.scala:231), which has no missing
parents
14/03/30 13:15:21 INFO DAGScheduler: Submitting 1 missing tasks from
Stage 165 (FlatMappedRDD[883] at flatMap at ALS.scala:231)
...
14/03/30 13:15:21 INFO SparkContext: Job finished: count at <console>:26,
took 0.030044 s
res21: Long = 1682
```

As expected, we have a factor array for each user (`943` factors) and movie (`1682` factors).

Training a model using implicit feedback data

The standard matrix factorization approach in MLlib deals with explicit ratings. To work with implicit data, you can use the `trainImplicit` method. It is called in a manner similar to the standard `train` method. There is an additional parameter, `alpha`, that can be set (and in the same way, the regularization parameter, `lambda`, should be selected via testing and cross-validation methods).

The `alpha` parameter controls the baseline level of confidence weighting applied. A higher level of `alpha` tends to make the model more confident about the fact that missing data equates to no preference for the relevant user-item pair.

As an exercise, try to take the existing MovieLens dataset and convert it into an implicit dataset. One possible approach is to convert it to binary feedback (0s and 1s) by applying a threshold on the ratings at some level.

Another approach could be to convert the ratings' values into confidence weights (for example, perhaps, low ratings could imply zero weights, or even negative weights, which are supported by MLlib's implementation).

Train a model on this dataset and compare the results of the following section with those generated by your implicit model.

Using the recommendation model

Now that we have our trained model, we're ready to use it to make predictions. These predictions typically take one of two forms: recommendations for a given user and related or similar items for a given item.

User recommendations

In this case, we would like to generate recommended items for a given user. This usually takes the form of a *top-K* list, that is, the *K* items that our model predicts will have the highest probability of the user liking them. This is done by computing the predicted score for each item and ranking the list based on this score.

The exact method to perform this computation depends on the model involved. For example, in user-based approaches, the ratings of similar users on items are used to compute the recommendations for a user, while in an item-based approach, the computation is based on the similarity of items the user has rated to the candidate items.

In matrix factorization, because we are modeling the ratings matrix directly, the predicted score can be computed as the vector dot product between a user-factor vector and an item-factor vector.

Generating movie recommendations from the MovieLens 100k dataset

As MLlib's recommendation model is based on matrix factorization, we can use the factor matrices computed by our model to compute predicted scores (or ratings) for a user. We will focus on the explicit rating case using MovieLens data; however, the approach is the same when using the implicit model.

The MatrixFactorizationModel class has a convenient predict method that will compute a predicted score for a given user and item combination:

```
val predictedRating = model.predict(789, 123)
```

The output is as follows:

```
14/03/30 16:10:10 INFO SparkContext: Starting job: lookup at MatrixFactorizationModel.scala:45
14/03/30 16:10:10 INFO DAGScheduler: Got job 30 (lookup at MatrixFactorizationModel.scala:45) with 1 output partitions (allowLocal=false)
...
```

```
14/03/30 16:10:10 INFO SparkContext: Job finished: lookup at
MatrixFactorizationModel.scala:46, took 0.023077 s
predictedRating: Double = 3.128545693368485
```

As we can see, this model predicts a rating of 3.12 for user 789 and movie 123.

 Note that you might see different results than those shown in this section because the ALS model is initialized randomly. So, different runs of the model will lead to different solutions.

The predict method can also take an RDD of (user, item) IDs as the input and will generate predictions for each of these. We can use this method to make predictions for many users and items at the same time.

To generate the *top-K* recommended items for a user, MatrixFactorizationModel provides a convenience method called recommendProducts. This takes two arguments: user and num, where user is the user ID, and num is the number of items to recommend.

It returns the top num items ranked in the order of the predicted score. Here, the scores are computed as the dot product between the user-factor vector and each item-factor vector.

Let's generate the top 10 recommended items for user 789:

```
val userId = 789
val K = 10
val topKRecs = model.recommendProducts(userId, K)
```

We now have a set of predicted ratings for each movie for user 789. If we print this out, we could inspect the top 10 recommendations for this user:

```
println(topKRecs.mkString("\n"))
```

You should see the following output on your console:

```
Rating(789,715,5.931851273771102)
Rating(789,12,5.582301095666215)
Rating(789,959,5.516272981542168)
Rating(789,42,5.458065302395629)
Rating(789,584,5.449949837103569)
Rating(789,750,5.348768847643657)
```

```
Rating(789,663,5.30832117499004)
Rating(789,134,5.278933936827717)
Rating(789,156,5.250959077906759)
Rating(789,432,5.169863417126231)
```

Inspecting the recommendations

We can give these recommendations a sense check by taking a quick look at the titles of the movies a user has rated and the recommended movies. First, we need to load the movie data (which is the one of the datasets we explored in the previous chapter). We'll collect this data as a `Map[Int, String]` method mapping the movie ID to the title:

```
val movies = sc.textFile("/PATH/ml-100k/u.item")
val titles = movies.map(line => line.split("\\|").take(2)).map(array
=> (array(0).toInt,
   array(1))).collectAsMap()
titles(123)
```

The preceding code will produce the output as:

`res68: String = Frighteners, The (1996)`

For our user `789`, we can find out what movies they have rated, take the 10 movies with the highest rating, and then check the titles. We will do this now by first using the `keyBy` Spark function to create an RDD of key-value pairs from our `ratings` RDD, where the key will be the user ID. We will then use the `lookup` function to return just the ratings for this key (that is, that particular user ID) to the driver:

```
val moviesForUser = ratings.keyBy(_.user).lookup(789)
```

Let's see how many movies this user has rated. This will be the `size` of the `moviesForUser` collection:

```
println(moviesForUser.size)
```

We will see that this user has rated `33` movies.

Next, we will take the 10 movies with the highest ratings by sorting the `moviesForUser` collection using the `rating` field of the `Rating` object. We will then extract the movie title for the relevant product ID attached to the `Rating` class from our mapping of movie titles and print out the top `10` titles with their ratings:

```
moviesForUser.sortBy(-_.rating).take(10).map(rating => (titles(rating.
product), rating.rating)).foreach(println)
```

You will see the following output displayed:

```
(Godfather, The (1972),5.0)
(Trainspotting (1996),5.0)
(Dead Man Walking (1995),5.0)
(Star Wars (1977),5.0)
(Swingers (1996),5.0)
(Leaving Las Vegas (1995),5.0)
(Bound (1996),5.0)
(Fargo (1996),5.0)
(Last Supper, The (1995),5.0)
(Private Parts (1997),4.0)
```

Now, let's take a look at the top 10 recommendations for this user and see what the titles are using the same approach as the one we used earlier (note that the recommendations are already sorted):

```
topKRecs.map(rating => (titles(rating.product), rating.rating)).
foreach(println)
```

The output is as follows:

```
(To Die For (1995),5.931851273771102)
(Usual Suspects, The (1995),5.582301095666215)
(Dazed and Confused (1993),5.516272981542168)
(Clerks (1994),5.458065302395629)
(Secret Garden, The (1993),5.449949837103569)
(Amistad (1997),5.348768847643657)
(Being There (1979),5.30832117499004)
(Citizen Kane (1941),5.278933936827717)
(Reservoir Dogs (1992),5.250959077906759)
(Fantasia (1940),5.169863417126231)
```

We leave it to you to decide whether these recommendations make sense.

Item recommendations

Item recommendations are about answering the following question: for a certain item, what are the items most similar to it? Here, the precise definition of similarity is dependent on the model involved. In most cases, similarity is computed by comparing the vector representation of two items using some similarity measure. Common similarity measures include Pearson correlation and cosine similarity for real-valued vectors and Jaccard similarity for binary vectors.

Generating similar movies for the MovieLens 100k dataset

The current `MatrixFactorizationModel` API does not directly support item-to-item similarity computations. Therefore, we will need to create our own code to do this.

We will use the cosine similarity metric, and we will use the jblas linear algebra library (a dependency of MLlib) to compute the required vector dot products. This is similar to how the existing `predict` and `recommendProducts` methods work, except that we will use cosine similarity as opposed to just the dot product.

We would like to compare the factor vector of our chosen item with each of the other items, using our similarity metric. In order to perform linear algebra computations, we will first need to create a vector object out of the factor vectors, which are in the form of an `Array[Double]`. The JBLAS class, `DoubleMatrix`, takes an `Array[Double]` as the constructor argument as follows:

```
import org.jblas.DoubleMatrix
val aMatrix = new DoubleMatrix(Array(1.0, 2.0, 3.0))
```

Here is the output of the preceding code:

aMatrix: org.jblas.DoubleMatrix = [1.000000; 2.000000; 3.000000]

> Note that using jblas, vectors are represented as a one-dimensional `DoubleMatrix` class, while matrices are a two-dimensional `DoubleMatrix` class.

We will need a method to compute the cosine similarity between two vectors. Cosine similarity is a measure of the angle between two vectors in an *n*-dimensional space. It is computed by first calculating the dot product between the vectors and then dividing the result by a denominator, which is the norm (or length) of each vector multiplied together (specifically, the L2-norm is used in cosine similarity). In this way, cosine similarity is a normalized dot product.

The cosine similarity measure takes on values between -1 and 1. A value of 1 implies completely similar, while a value of 0 implies independence (that is, no similarity). This measure is useful because it also captures negative similarity, that is, a value of -1 implies that not only are the vectors not similar, but they are also completely dissimilar.

Let's create our `cosineSimilarity` function here:

```
def cosineSimilarity(vec1: DoubleMatrix, vec2: DoubleMatrix): Double = 
{
  vec1.dot(vec2) / (vec1.norm2() * vec2.norm2())
}
```

> Note that we defined a return type for this function of `Double`. We are not required to do this, since Scala features type inference. However, it can often be useful to document return types for Scala functions.

Let's try it out on one of our item factors for item `567`. We will need to collect an item factor from our model; we will do this using the `lookup` method in a similar way that we did earlier to collect the ratings for a specific user. In the following lines of code, we also use the `head` function, since `lookup` returns an array of values, and we only need the first value (in fact, there will only be one value, which is the factor vector for this item).

Since this will be an `Array[Double]`, we will then need to create a `DoubleMatrix` object from it and compute the cosine similarity with itself:

```
val itemId = 567
val itemFactor = model.productFeatures.lookup(itemId).head
val itemVector = new DoubleMatrix(itemFactor)
cosineSimilarity(itemVector, itemVector)
```

A similarity metric should measure how close, in some sense, two vectors are to each other. Here, we can see that our cosine similarity metric tells us that this item vector is identical to itself, which is what we would expect:

res113: Double = 1.0

Now, we are ready to apply our similarity metric to each item:

```
val sims = model.productFeatures.map{ case (id, factor) =>
  val factorVector = new DoubleMatrix(factor)
  val sim = cosineSimilarity(factorVector, itemVector)
  (id, sim)
}
```

Next, we can compute the top 10 most similar items by sorting out the similarity score for each item:

```
// recall we defined K = 10 earlier
val sortedSims = sims.top(K)(Ordering.by[(Int, Double), Double] { case
(id, similarity) => similarity })
```

In the preceding code snippet, we used Spark's `top` function, which is an efficient way to compute *top-K* results in a distributed fashion, instead of using `collect` to return all the data to the driver and sorting it locally (remember that we could be dealing with millions of users and items in the case of recommendation models).

We need to tell Spark how to sort the (item id, similarity score) pairs in the sims RDD. To do this, we will pass an extra argument to top, which is a Scala Ordering object that tells Spark that it should sort by the value in the key-value pair (that is, sort by similarity).

Finally, we can print the 10 items with the highest computed similarity metric to our given item:

```
println(sortedSims.take(10).mkString("\n"))
```

You will see output like the following one:

(567,1.0000000000000002)
(1471,0.6932331537649621)
(670,0.6898690594544726)
(201,0.6897964975027041)
(343,0.6891221044611473)
(563,0.6864214133620066)
(294,0.6812075443259535)
(413,0.6754663844488256)
(184,0.6702643811753909)
(109,0.6594872765176396)

Not surprisingly, we can see that the top-ranked similar item is our item. The rest are the other items in our set of items, ranked in order of our similarity metric.

Inspecting the similar items

Let's see what the title of our chosen movie is:

```
println(titles(itemId))
```

The preceding code will print the following output:

Wes Craven's New Nightmare (1994)

As we did for user recommendations, we can sense check our item-to-item similarity computations and take a look at the titles of the most similar movies. This time, we will take the top 11 so that we can exclude our given movie. So, we will take the numbers 1 to 11 in the list:

```
val sortedSims2 = sims.top(K + 1)(Ordering.by[(Int, Double), Double] {
  case (id, similarity) => similarity })
sortedSims2.slice(1, 11).map{ case (id, sim) => (titles(id), sim)
}.mkString("\n")
```

You will see the movie titles and scores displayed similar to this output:

```
(Hideaway (1995),0.6932331537649621)
(Body Snatchers (1993),0.6898690594544726)
(Evil Dead II (1987),0.6897964975027041)
(Alien: Resurrection (1997),0.6891221044611473)
(Stephen King's The Langoliers (1995),0.6864214133620066)
(Liar Liar (1997),0.6812075443259535)
(Tales from the Crypt Presents: Bordello of Blood
(1996),0.6754663844488256)
(Army of Darkness (1993),0.6702643811753909)
(Mystery Science Theater 3000: The Movie (1996),0.6594872765176396)
(Scream (1996),0.6538249646863378)
```

Once again note that you might see quite different results due to random model initialization.

Now that you have computed similar items using cosine similarity, see if you can do the same with the user-factor vectors to compute similar users for a given user.

Evaluating the performance of recommendation models

How do we know whether the model we have trained is a good model? We need to be able to evaluate its predictive performance in some way. **Evaluation metrics** are measures of a model's predictive capability or accuracy. Some are direct measures of how well a model predicts the model's target variable (such as Mean Squared Error), while others are concerned with how well the model performs at predicting things that might not be directly optimized in the model but are often closer to what we care about in the real world (such as Mean average precision).

Evaluation metrics provide a standardized way of comparing the performance of the same model with different parameter settings and of comparing performance across different models. Using these metrics, we can perform model selection to choose the best-performing model from the set of models we wish to evaluate.

Here, we will show you how to calculate two common evaluation metrics used in recommender systems and collaborative filtering models: Mean Squared Error and Mean average precision at K.

Mean Squared Error

The **Mean Squared Error** (**MSE**) is a direct measure of the reconstruction error of the user-item rating matrix. It is also the objective function being minimized in certain models, specifically many matrix-factorization techniques, including ALS. As such, it is commonly used in explicit ratings settings.

It is defined as the sum of the squared errors divided by the number of observations. The squared error, in turn, is the square of the difference between the predicted rating for a given user-item pair and the actual rating.

We will use our user 789 as an example. Let's take the first rating for this user from the moviesForUser set of Ratings that we previously computed:

```
val actualRating = moviesForUser.take(1)(0)
```

Here is the output:

```
actualRating: org.apache.spark.mllib.recommendation.Rating =
Rating(789,1012,4.0)
```

We will see that the rating for this user-item combination is 4. Next, we will compute the model's predicted rating:

```
val predictedRating = model.predict(789, actualRating.product)
```

The output of the model's predicted rating is as follows:

```
...
14/04/13 13:01:15 INFO SparkContext: Job finished: lookup at
MatrixFactorizationModel.scala:46, took 0.025404 s

predictedRating: Double = 4.001005374200248
```

We will see that the predicted rating is about 4, very close to the actual rating. Finally, we will compute the squared error between the actual rating and the predicted rating:

```
val squaredError = math.pow(predictedRating - actualRating.rating,
  2.0)
```

The preceding code will output the squared error:

```
squaredError: Double = 1.010777282523947E-6
```

So, in order to compute the overall MSE for the dataset, we need to compute this squared error for each (user, movie, actual rating, predicted rating) entry, sum them up, and divide them by the number of ratings. We will do this in the following code snippet.

> Note the following code is adapted from the Apache Spark programming guide for ALS at http://spark.apache.org/docs/latest/mllib-collaborative-filtering.html.

First, we will extract the user and product IDs from the `ratings` RDD and make predictions for each user-item pair using `model.predict`. We will use the user-item pair as the key and the predicted rating as the value:

```
val usersProducts = ratings.map{ case Rating(user, product, rating)
  => (user, product)}
val predictions = model.predict(usersProducts).map{
    case Rating(user, product, rating) => ((user, product), rating)
}
```

Next, we extract the actual ratings and also map the `ratings` RDD so that the user-item pair is the key and the actual rating is the value. Now that we have two RDDs with the same form of key, we can join them together to create a new RDD with the actual and predicted ratings for each user-item combination:

```
val ratingsAndPredictions = ratings.map{
  case Rating(user, product, rating) => ((user, product), rating)
}.join(predictions)
```

Finally, we will compute the MSE by summing up the squared errors using `reduce` and dividing by the `count` method of the number of records:

```
val MSE = ratingsAndPredictions.map{
    case ((user, product), (actual, predicted)) =>  math.pow((actual - predicted), 2)
}.reduce(_ + _) / ratingsAndPredictions.count
println("Mean Squared Error = " + MSE)
```

The output is as follows:

Mean Squared Error = 0.08231947642632852

It is common to use the **Root Mean Squared Error** (**RMSE**), which is just the square root of the MSE metric. This is somewhat more interpretable, as it is in the same units as the underlying data (that is, the ratings in this case). It is equivalent to the standard deviation of the differences between the predicted and actual ratings. We can compute it simply as follows:

```
val RMSE = math.sqrt(MSE)
println("Root Mean Squared Error = " + RMSE)
```

The preceding code will print the Root Mean Squared Error:

Root Mean Squared Error = 0.2869137090247319

Mean average precision at K

Mean average precision at K (MAPK) is the mean of the **average precision at K (APK)** metric across all instances in the dataset. APK is a metric commonly used in information retrieval. APK is a measure of the average relevance scores of a set of the *top-K* documents presented in response to a query. For each query instance, we will compare the set of *top-K* results with the set of actual relevant documents (that is, a ground truth set of relevant documents for the query).

In the APK metric, the order of the result set matters, in that, the APK score would be higher if the result documents are both relevant and the relevant documents are presented higher in the results. It is, thus, a good metric for recommender systems in that typically we would compute the *top-K* recommended items for each user and present these to the user. Of course, we prefer models where the items with the highest predicted scores (which are presented at the top of the list of recommendations) are, in fact, the most relevant items for the user. APK and other ranking-based metrics are also more appropriate evaluation measures for implicit datasets; here, MSE makes less sense.

In order to evaluate our model, we can use APK, where each user is the equivalent of a query, and the set of *top-K* recommended items is the document result set. The relevant documents (that is, the ground truth) in this case, is the set of items that a user interacted with. Hence, APK attempts to measure how good our model is at predicting items that a user will find relevant and choose to interact with.

> The code for the following average precision computation is based on https://github.com/benhamner/Metrics.
>
> More information on MAPK can be found at https://www.kaggle.com/wiki/MeanAveragePrecision.

Our function to compute the APK is shown here:

```
def avgPrecisionK(actual: Seq[Int], predicted: Seq[Int], k: Int):
Double = {
  val predK = predicted.take(k)
  var score = 0.0
  var numHits = 0.0
  for ((p, i) <- predK.zipWithIndex) {
    if (actual.contains(p)) {
      numHits += 1.0
      score += numHits / (i.toDouble + 1.0)
    }
  }
  if (actual.isEmpty) {
    1.0
```

```
    } else {
      score / scala.math.min(actual.size, k).toDouble
    }
  }
```

As you can see, this takes as input a list of `actual` item IDs that are associated with the user and another list of `predicted` ids so that our estimate will be relevant for the user.

We can compute the APK metric for our example user `789` as follows. First, we will extract the actual movie IDs for the user:

```
val actualMovies = moviesForUser.map(_.product)
```

The output is as follows:

```
actualMovies: Seq[Int] = ArrayBuffer(1012, 127, 475, 93, 1161, 286, 293,
9, 50, 294, 181, 1, 1008, 508, 284, 1017, 137, 111, 742, 248, 249, 1007,
591, 150, 276, 151, 129, 100, 741, 288, 762, 628, 124)
```

We will then use the movie recommendations we made previously to compute the APK score using K = 10:

```
val predictedMovies = topKRecs.map(_.product)
```

Here is the output:

```
predictedMovies: Array[Int] = Array(27, 497, 633, 827, 602, 849, 401,
584, 1035, 1014)
```

The following code will produce the average precision:

```
val apk10 = avgPrecisionK(actualMovies, predictedMovies, 10)
```

The preceding code will print:

```
apk10: Double = 0.0
```

In this case, we can see that our model is not doing a very good job of predicting relevant movies for this user as the APK score is 0.

In order to compute the APK for each user and average them to compute the overall MAPK, we will need to generate the list of recommendations for each user in our dataset. While this can be fairly intensive on a large scale, we can distribute the computation using our Spark functionality. However, one limitation is that each worker must have the full item-factor matrix available so that it can compute the dot product between the relevant user vector and all item vectors. This can be a problem when the number of items is extremely high as the item matrix must fit in the memory of one machine.

> There is actually no easy way around this limitation. One possible approach is to only compute recommendations for a subset of items from the total item set, using approximate techniques such as Locality Sensitive Hashing (http://en.wikipedia.org/wiki/Locality-sensitive_hashing).

We will now see how to go about this. First, we will collect the item factors and form a `DoubleMatrix` object from them:

```
val itemFactors = model.productFeatures.map { case (id, factor) =>
factor }.collect()
val itemMatrix = new DoubleMatrix(itemFactors)
println(itemMatrix.rows, itemMatrix.columns)
```

The output of the preceding code is as follows:

`(1682,50)`

This gives us a matrix with `1682` rows and `50` columns, as we would expect from `1682` movies with a factor dimension of `50`. Next, we will distribute the item matrix as a broadcast variable so that it is available on each worker node:

```
val imBroadcast = sc.broadcast(itemMatrix)
```

You will see the output as follows:

`14/04/13 21:02:01 INFO MemoryStore: ensureFreeSpace(672960) called with curMem=4006896, maxMem=311387750`

`14/04/13 21:02:01 INFO MemoryStore: Block broadcast_21 stored as values to memory (estimated size 657.2 KB, free 292.5 MB)`

`imBroadcast: org.apache.spark.broadcast.Broadcast[org.jblas.DoubleMatrix] = Broadcast(21)`

Now we are ready to compute the recommendations for each user. We will do this by applying a `map` function to each user factor within which we will perform a matrix multiplication between the user-factor vector and the movie-factor matrix. The result is a vector (of length `1682`, that is, the number of movies we have) with the predicted rating for each movie. We will then sort these predictions by the predicted rating:

```
val allRecs = model.userFeatures.map{ case (userId, array) =>
  val userVector = new DoubleMatrix(array)
  val scores = imBroadcast.value.mmul(userVector)
  val sortedWithId = scores.data.zipWithIndex.sortBy(-_._1)
  val recommendedIds = sortedWithId.map(_._2 + 1).toSeq
  (userId, recommendedIds)
}
```

You will see the following on the screen:

```
allRecs: org.apache.spark.rdd.RDD[(Int, Seq[Int])] = MappedRDD[269] at map at <console>:29
```

As we can see, we now have an RDD that contains a list of movie IDs for each user ID. These movie IDs are sorted in order of the estimated rating.

 Note that we needed to add 1 to the returned movie ids (as highlighted in the preceding code snippet), as the item-factor matrix is 0-indexed, while our movie IDs start at 1.

We also need the list of movie IDs for each user to pass into our APK function as the `actual` argument. We already have the `ratings` RDD ready, so we can extract just the user and movie IDs from it.

If we use Spark's `groupBy` operator, we will get an RDD that contains a list of (userid, movieid) pairs for each user ID (as the user ID is the key on which we perform the `groupBy` operation):

```
val userMovies = ratings.map{ case Rating(user, product, rating) => (user, product) }.groupBy(_._1)
```

The output of the preceding code is as follows:

```
userMovies: org.apache.spark.rdd.RDD[(Int, Seq[(Int, Int)])] = MapPartitionsRDD[277] at groupBy at <console>:21
```

Finally, we can use Spark's `join` operator to join these two RDDs together on the user ID key. Then, for each user, we have the list of actual and predicted movie IDs that we can pass to our APK function. In a manner similar to how we computed MSE, we will sum each of these APK scores using a `reduce` action and divide by the number of users (that is, the count of the `allRecs` RDD):

```
val K = 10
val MAPK = allRecs.join(userMovies).map{ case (userId, (predicted, actualWithIds)) =>
  val actual = actualWithIds.map(_._2).toSeq
  avgPrecisionK(actual, predicted, K)
}.reduce(_ + _) / allRecs.count
println("Mean Average Precision at K = " + MAPK)
```

The preceding code will print the mean average precision at K as follows:

```
Mean Average Precision at K = 0.030486963254725705
```

Our model achieves a fairly low MAPK. However, note that typical values for recommendation tasks are usually relatively low, especially if the item set is extremely large.

Try out a few parameter settings for `lambda` and `rank` (and `alpha` if you are using the implicit version of ALS) and see whether you can find a model that performs better based on the RMSE and MAPK evaluation metrics.

Using MLlib's built-in evaluation functions

While we have computed MSE, RMSE, and MAPK from scratch, and it a useful learning exercise to do so, MLlib provides convenience functions to do this for us in the `RegressionMetrics` and `RankingMetrics` classes.

RMSE and MSE

First, we will compute the MSE and RMSE metrics using `RegressionMetrics`. We will instantiate a `RegressionMetrics` instance by passing in an RDD of key-value pairs that represent the predicted and true values for each data point, as shown in the following code snippet. Here, we will again use the `ratingsAndPredictions` RDD we computed in our earlier example:

```
import org.apache.spark.mllib.evaluation.RegressionMetrics
val predictedAndTrue = ratingsAndPredictions.map { case ((user, product), (predicted, actual)) => (predicted, actual) }
val regressionMetrics = new RegressionMetrics(predictedAndTrue)
```

We can then access various metrics, including MSE and RMSE. We will print out these metrics here:

```
println("Mean Squared Error = " + regressionMetrics.meanSquaredError)
println("Root Mean Squared Error = " + regressionMetrics.rootMeanSquaredError)
```

You will see that the output for MSE and RMSE is exactly the same as the metrics we computed earlier:

```
Mean Squared Error = 0.08231947642632852
Root Mean Squared Error = 0.2869137090247319
```

MAP

As we did for MSE and RMSE, we can compute ranking-based evaluation metrics using MLlib's `RankingMetrics` class. Similarly, to our own average precision function, we need to pass in an RDD of key-value pairs, where the key is an `Array` of predicted item IDs for a user, while the value is an array of actual item IDs.

Building a Recommendation Engine with Spark

The implementation of the average precision at the K function in `RankingMetrics` is slightly different from ours, so we will get different results. However, the computation of the overall mean average precision (MAP, which does not use a threshold at K) is the same as our function if we select K to be very high (say, at least as high as the number of items in our item set):

First, we will calculate MAP using `RankingMetrics`:

```
import org.apache.spark.mllib.evaluation.RankingMetrics
val predictedAndTrueForRanking = allRecs.join(userMovies).map{ case
(userId, (predicted, actualWithIds)) =>
  val actual = actualWithIds.map(_._2)
  (predicted.toArray, actual.toArray)
}
val rankingMetrics = new RankingMetrics(predictedAndTrueForRanking)
println("Mean Average Precision = " + rankingMetrics.
meanAveragePrecision)
```

You will see the following output:

Mean Average Precision = 0.07171412913757183

Next, we will use our function to compute the MAP in exactly the same way as we did previously, except that we set K to a very high value, say `2000`:

```
val MAPK2000 = allRecs.join(userMovies).map{ case (userId, (predicted,
actualWithIds)) =>
  val actual = actualWithIds.map(_._2).toSeq
  avgPrecisionK(actual, predicted, 2000)
}.reduce(_ + _) / allRecs.count
println("Mean Average Precision = " + MAPK2000)
```

You will see that the MAP from our own function is the same as the one computed using `RankingMetrics`:

Mean Average Precision = 0.07171412913757186

> We will not cover cross-validation in this chapter, as we will provide a detailed treatment in the next few chapters. However, note that the same techniques for cross-validation that are explored in the upcoming chapters can be used to evaluate recommendation models, using the performance metrics such as MSE, RMSE, and MAP, which we covered in this section.

Summary

In this chapter, we used Spark's MLlib library to train a collaborative filtering recommendation model, and you learned how to use this model to make predictions for the items that a given user might have a preference for. We also used our model to find items that are similar or related to a given item. Finally, we explored common metrics to evaluate the predictive capability of our recommendation model.

In the next chapter, you will learn how to use Spark to train a model to classify your data and to use standard evaluation mechanisms to gauge the performance of your model.

5
Building a Classification Model with Spark

In this chapter, you will learn the basics of classification models and how they can be used in a variety of contexts. Classification generically refers to classifying things into distinct categories or classes. In the case of a classification model, we typically wish to assign classes based on a set of features. The features might represent variables related to an item or object, an event or context, or some combination of these.

The simplest form of classification is when we have two classes; this is referred to as binary classification. One of the classes is usually labeled as the positive class (assigned a label of 1), while the other is labeled as the negative class (assigned a label of -1 or, sometimes, 0).

A simple example with two classes is shown in the following figure. The input features in this case have two dimensions, and the feature values are represented on the x and y axes in the figure.

Building a Classification Model with Spark

Our task is to train a model that can classify new data points in this two-dimensional space as either one class (red) or the other (blue).

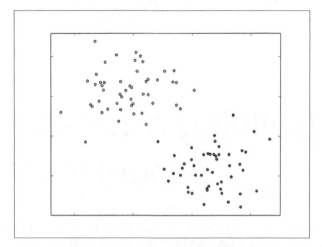

A simple binary classification problem

If we have more than two classes, we would refer to multiclass classification, and classes are typically labeled using integer numbers starting at 0 (for example, five different classes would range from label 0 to 4). An example is shown in the following figure. Again, the input features are assumed to be two-dimensional for ease of illustration.

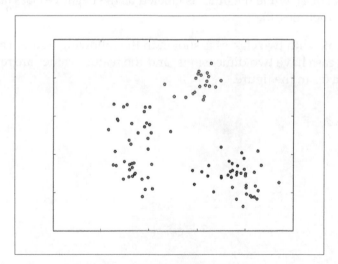

A simple multiclass classification problem

Classification is a form of supervised learning where we train a model with training examples that include known targets or outcomes of interest (that is, the model is supervised with these example outcomes). Classification models can be used in many situations, but a few common examples include:

- Predicting the probability of Internet users clicking on an online advert; here, the classes are binary in nature (that is, click or no click)
- Detecting fraud; again, in this case, the classes are commonly binary (fraud or no fraud)
- Predicting defaults on loans (binary)
- Classifying images, video, or sounds (most often multiclass, with potentially very many different classes)
- Assigning categories or tags to news articles, web pages, or other content (multiclass)
- Discovering e-mail and web spam, network intrusions, and other malicious behavior (binary or multiclass)
- Detecting failure situations, for example in computer systems or networks
- Ranking customers or users in order of probability that they might purchase a product or use a service (this can be framed as classification by predicting probabilities and then ranking in the descending order)
- Predicting customers or users who might stop using a product, service, or provider (called churn)

These are just a few possible use cases. In fact, it is probably safe to say that classification is one of the most widely used machine learning and statistical techniques in modern businesses and especially online businesses.

In this chapter, we will:

- Discuss the types of classification models available in MLlib
- Use Spark to extract the appropriate features from raw input data
- Train a number of classification models using MLlib
- Make predictions with our classification models
- Apply a number of standard evaluation techniques to assess the predictive performance of our models
- Illustrate how to improve model performance using some of the feature-extraction approaches from *Chapter 3, Obtaining, Processing, and Preparing Data with Spark*
- Explore the impact of parameter tuning on model performance and learn how to use cross-validation to select the most optimal model parameters

Types of classification models

We will explore three common classification models available in Spark: linear models, decision trees, and naïve Bayes models. Linear models, while less complex, are relatively easier to scale to very large datasets. Decision tree is a powerful nonlinear technique that can be a little more difficult to scale up (fortunately, MLlib takes care of this for us!) and more computationally intensive to train, but delivers leading performance in many situations. Naïve Bayes models are more simple but are easy to train efficiently and parallelize (in fact, they require only one pass over the dataset). They can also give reasonable performance in many cases when appropriate feature engineering is used. A naïve Bayes model also provides a good baseline model against which we can measure the performance of other models.

Currently, Spark's MLlib library supports binary classification for linear models, decision trees, and naïve Bayes models and multiclass classification for decision trees and naïve Bayes models. In this book, for simplicity in illustrating the examples, we will focus on the binary case.

Linear models

The core idea of linear models (or generalized linear models) is that we model the predicted outcome of interest (often called the target or dependent variable) as a function of a simple linear predictor applied to the input variables (also referred to as features or independent variables).

$$y = f(w^T x)$$

Here, y is the target variable, w is the vector of parameters (known as the weight vector), and x is the vector of input features.

$w^T x$ is the linear predictor (or vector dot product) of the weight vector w and feature vector x. To this linear predictor, we applied a function f (called the link function).

Linear models can, in fact, be used for both classification and regression, simply by changing the link function. Standard linear regression (covered in the next chapter) uses an identity link (that is, $y = w^T x$ directly), while binary classification uses alternative link functions as discussed here.

Let's take a look at the example of online advertising. In this case, the target variable would be 0 (often assigned the class label of -1 in mathematical treatments) if no click was observed for a given advert displayed on a web page (called an impression). The target variable would be 1 if a click occurred. The feature vector for each impression would consist of variables related to the impression event (such as features relating to the user, web page, advert and advertiser, and various other factors relating to the context of the event, such as the type of device used, time of the day, and geolocation).

Thus, we would like to find a model that maps a given input feature vector (advert impression) to a predicted outcome (click or not). To make a prediction for a new data point, we will take the new feature vector (which is unseen, and hence, we do not know what the target variable is) and compute the dot product with our weight vector. We will then apply the relevant link function, and the result is our predicted outcome (after applying a threshold to the prediction, in the case of some models).

Given a set of input data in the form of feature vectors and target variables, we would like to find the weight vector that is the best fit for the data, in the sense that we minimize some error between what our model predicts and the actual outcomes observed. This process is called **model fitting, training,** or **optimization**.

More formally, we seek to find the weight vector that minimizes the sum, over all the training examples, of the loss (or error) computed from some loss function. The loss function takes the weight vector, feature vector, and the actual outcome for a given training example as input and outputs the loss. In fact, the loss function itself is effectively specified by the link function; hence, for a given type of classification or regression (that is, a given link function), there is a corresponding loss function.

>
> For further details on linear models and loss functions, see the linear methods section related to binary classification in the *Spark Programming Guide* at http://spark.apache.org/docs/latest/mllib-linear-methods.html#binary-classification.
>
> Also, see the Wikipedia entry for generalized linear models at http://en.wikipedia.org/wiki/Generalized_linear_model.

While a detailed treatment of linear models and loss functions is beyond the scope of this book, MLlib provides two loss functions suitable to binary classification (you can learn more about them from the Spark documentation). The first one is logistic loss, which equates to a model known as **logistic regression**, while the second one is the hinge loss, which is equivalent to a linear **Support Vector Machine** (**SVM**). Note that the SVM does not strictly fall into the statistical framework of generalized linear models but can be used in the same way as it essentially specifies a loss and link function.

In the following image, we show the logistic loss and hinge loss relative to the actual zero-one loss. The zero-one loss is the true loss for binary classification—it is either zero if the model predicts correctly or one if the model predicts incorrectly. The reason it is not actually used is that it is not a differentiable loss function, so it is not possible to easily compute a gradient and, thus, very difficult to optimize.

The other loss functions are approximations to the zero-one loss that make optimization possible.

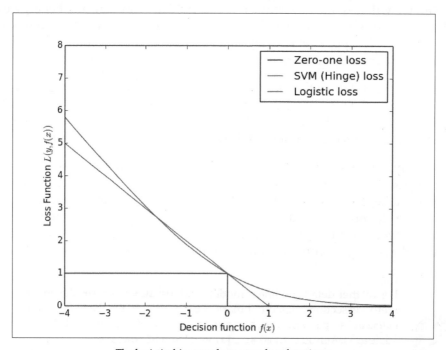

The logistic, hinge and zero-one loss functions

 The preceding loss diagram is adapted from the scikit-learn example at http://scikit-learn.org/stable/auto_examples/linear_model/plot_sgd_loss_functions.html.

Logistic regression

Logistic regression is a probabilistic model—that is, its predictions are bounded between 0 and 1, and for binary classification equate to the model's estimate of the probability of the data point belonging to the positive class. Logistic regression is one of the most widely used linear classification models.

As mentioned earlier, the link function used in logistic regression is the logit link:

$$1 / (1 + \exp(-w^T x))$$

The related loss function for logistic regression is the logistic loss:

$$\log(1 + \exp(-yw^Tx))$$

Here, y is the actual target variable (either *1* for the positive class or *-1* for the negative class).

Linear support vector machines

SVM is a powerful and popular technique for regression and classification. Unlike logistic regression, it is not a probabilistic model but predicts classes based on whether the model evaluation is positive or negative.

The SVM link function is the identity link, so the predicted outcome is:

$$y = w^Tx$$

Hence, if the evaluation of w^Tx is greater than or equal to a threshold of 0, the SVM will assign the data point to class 1; otherwise, the SVM will assign it to class 0 (this threshold is a model parameter of SVM and can be adjusted).

The loss function for SVM is known as the **hinge loss** and is defined as:

$$\max(0, 1 - yw^Tx)$$

SVM is a maximum margin classifier — it tries to find a weight vector such that the classes are separated as much as possible. It has been shown to perform well on many classification tasks, and the linear variant can scale to very large datasets.

> SVMs have a large amount of theory behind them, which is beyond the scope of this book, but you can visit `http://en.wikipedia.org/wiki/Support_vector_machine` and `http://www.support-vector-machines.org/` for more details.

In the following image, we have plotted the different decision functions for logistic regression (the blue line) and linear SVM (the red line), based on the simple binary classification example explained earlier.

You can see that the SVM effectively focuses on the points that lie closest to the decision function (the margin lines are shown with red dashes):

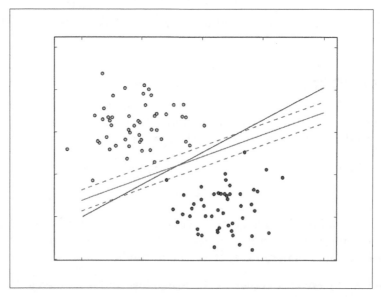

Decision functions for logistic regression and linear SVM for binary classification

The naïve Bayes model

Naïve Bayes is a probabilistic model that makes predictions by computing the probability of a data point that belongs to a given class. A naïve Bayes model assumes that each feature makes an independent contribution to the probability assigned to a class (it assumes conditional independence between features).

Due to this assumption, the probability of each class becomes a function of the product of the probability of a feature occurring, given the class, as well as the probability of this class. This makes training the model tractable and relatively straightforward. The class prior probabilities and feature conditional probabilities are all estimated from the frequencies present in the dataset. Classification is performed by selecting the most probable class, given the features and class probabilities.

An assumption is also made about the feature distributions (the parameters of which are estimated from the data). MLlib implements multinomial naïve Bayes that assumes that the feature distribution is a multinomial distribution that represents non-negative frequency counts of the features.

It is suitable for binary features (for example, *1-of-k* encoded categorical features) and is commonly used for text and document classification (where, as we have seen in *Chapter 3, Obtaining, Processing, and Preparing Data with Spark*, the bag-of-words vector is a typical feature representation).

> Take a look at the *MLlib - Naive Bayes* section in the Spark documentation at `http://spark.apache.org/docs/latest/mllib-naive-bayes.html` for more information.
>
> The Wikipedia page at `http://en.wikipedia.org/wiki/Naive_Bayes_classifier` has a more detailed explanation of the mathematical formulation.

Here, we have shown the decision function of naïve Bayes on our simple binary classification example:

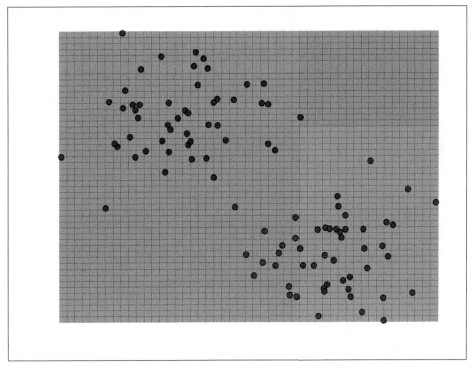

Decision function of naïve Bayes for binary classification

Decision trees

Decision tree model is a powerful, nonprobabilistic technique that can capture more complex nonlinear patterns and feature interactions. They have been shown to perform well on many tasks, are relatively easy to understand and interpret, can handle categorical and numerical features, and do not require input data to be scaled or standardized. They are well suited to be included in ensemble methods (for example, ensembles of decision tree models, which are called decision forests).

The decision tree model constructs a tree where the leaves represent a class assignment to class 0 or 1, and the branches are a set of features. In the following figure, we show a simple decision tree where the binary outcome is **Stay at home** or **Go to the beach**. The features are the weather outside.

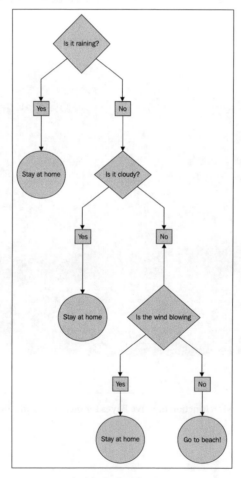

A simple decision tree

The decision tree algorithm is a top-down approach that begins at a root node (or feature), and then selects a feature at each step that gives the best split of the dataset, as measured by the information gain of this split. The information gain is computed from the node impurity (which is the extent to which the labels at the node are similar, or homogenous) minus the weighted sum of the impurities for the two child nodes that would be created by the split. For classification tasks, there are two measures that can be used to select the best split. These are Gini impurity and entropy.

> See the *MLlib - Decision Tree* section in the *Spark Programming Guide* at http://spark.apache.org/docs/latest/mllib-decision-tree.html for further details on the decision tree algorithm and impurity measures for classification.

In the following screenshot, we have plotted the decision boundary for the decision tree model, as we did for the other models earlier. We can see that the decision tree is able to fit complex, nonlinear models.

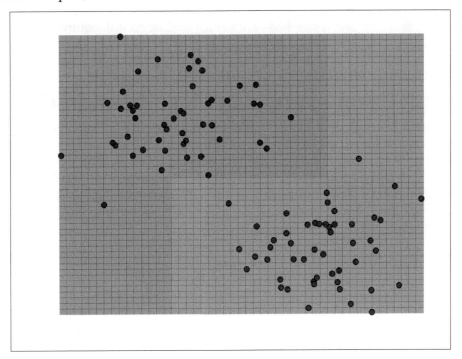

Decision function for a decision tree for binary classification

Extracting the right features from your data

You might recall from *Chapter 3, Obtaining, Processing, and Preparing Data with Spark* that the majority of machine learning models operate on numerical data in the form of feature vectors. In addition, for supervised learning methods such as classification and regression, we need to provide the target variable (or variables in the case of multiclass situations) together with the feature vector.

Classification models in MLlib operate on instances of `LabeledPoint`, which is a wrapper around the target variable (called the **label**) and the **feature vector**:

```
case class LabeledPoint(label: Double, features: Vector)
```

While in most examples of using classification, you will come across existing datasets that are already in the vector format, in practice, you will usually start with raw data that needs to be transformed into features. As we have already seen, this can involve preprocessing and transformation, such as binning numerical features, scaling and normalizing features, and using *1-of-k* encodings for categorical features.

Extracting features from the Kaggle/ StumbleUpon evergreen classification dataset

In this chapter, we will use a different dataset from the one we used for our recommendation model, as the MovieLens data doesn't have much for us to work with in terms of a classification problem. We will use a dataset from a competition on Kaggle. The dataset was provided by StumbleUpon, and the problem relates to classifying whether a given web page is ephemeral (that is, short lived and will cease being popular soon) or evergreen (that is, persistently popular) on their web content recommendation pages.

> The dataset used here can be downloaded from http://www.kaggle.com/c/stumbleupon/data.
> Download the training data (`train.tsv`) — you will need to accept the terms and conditions before downloading the dataset.
> You can find more information about the competition at http://www.kaggle.com/c/stumbleupon.

Before we begin, it will be easier for us to work with the data in Spark if we remove the column name header from the first line of the file. Change to the directory in which you downloaded the data (referred to as PATH here) and run the following command to remove the first line and pipe the result to a new file called train_noheader.tsv:

```
>sed 1d train.tsv > train_noheader.tsv
```

Now, we are ready to start up our Spark shell (remember to run this command from your Spark installation directory):

```
>./bin/spark-shell --driver-memory 4g
```

You can type in the code that follows for the remainder of this chapter directly into your Spark shell.

In a manner similar to what we did in the earlier chapters, we will load the raw training data into an RDD and inspect it:

```
val rawData = sc.textFile("/PATH/train_noheader.tsv")
val records = rawData.map(line => line.split("\t"))
records.first()
```

You will the following on the screen:

```
Array[String] = Array("http://www.bloomberg.com/news/2010-12-23/ibm-predicts-holographic-calls-air-breathing-batteries-by-2015.html", "4042",
...
```

You can check the fields that are available by reading through the overview on the dataset page above. The first two columns contain the URL and ID of the page. The next column contains some raw textual content. The next column contains the category assigned to the page. The next 22 columns contain numeric or categorical features of various kinds. The final column contains the target—1 is evergreen, while 0 is non-evergreen.

We'll start off with a simple approach of using only the available numeric features directly. As each categorical variable is binary, we already have a *1-of-k* encoding for these variables, so we don't need to do any further feature extraction.

Due to the way the data is formatted, we will have to do a bit of data cleaning during our initial processing by trimming out the extra quotation characters ("). There are also missing values in the dataset; they are denoted by the "?" character. In this case, we will simply assign a zero value to these missing values:

```
import org.apache.spark.mllib.regression.LabeledPoint
import org.apache.spark.mllib.linalg.Vectors
val data = records.map { r =>
```

```
    val trimmed = r.map(_.replaceAll("\"", ""))
    val label = trimmed(r.size - 1).toInt
    val features = trimmed.slice(4, r.size - 1).map(d => if (d ==
    "?") 0.0 else d.toDouble)
    LabeledPoint(label, Vectors.dense(features))
}
```

In the preceding code, we extracted the label variable from the last column and an array of features for columns 5 to 25 after cleaning and dealing with missing values. We converted the label to an `Int` value and the features to an `Array[Double]`. Finally, we wrapped the label and features in a `LabeledPoint` instance, converting the features into an MLlib `Vector`.

We will also cache the data and count the number of data points:

```
data.cache
val numData = data.count
```

You will see that the value of `numData` is 7395.

We will explore the dataset in more detail a little later, but we will tell you now that there are some negative feature values in the numeric data. As we saw earlier, the naïve Bayes model requires non-negative features and will throw an error if it encounters negative values. So, for now, we will create a version of our input feature vectors for the naïve Bayes model by setting any negative feature values to zero:

```
val nbData = records.map { r =>
    val trimmed = r.map(_.replaceAll("\"", ""))
    val label = trimmed(r.size - 1).toInt
    val features = trimmed.slice(4, r.size - 1).map(d => if (d ==
    "?") 0.0 else d.toDouble).map(d => if (d < 0) 0.0 else d)
    LabeledPoint(label, Vectors.dense(features))
}
```

Training classification models

Now that we have extracted some basic features from our dataset and created our input RDD, we are ready to train a number of models. To compare the performance and use of different models, we will train a model using logistic regression, SVM, naïve Bayes, and a decision tree. You will notice that training each model looks nearly identical, although each has its own specific model parameters that can be set. MLlib sets sensible defaults in most cases, but in practice, the best parameter setting should be selected using evaluation techniques, which we will cover later in this chapter.

Training a classification model on the Kaggle/StumbleUpon evergreen classification dataset

We can now apply the models from MLlib to our input data. First, we need to import the required classes and set up some minimal input parameters for each model. For logistic regression and SVM, this is the number of iterations, while for the decision tree model, it is the maximum tree depth:

```
import org.apache.spark.mllib.classification.LogisticRegressionWithSGD
import org.apache.spark.mllib.classification.SVMWithSGD
import org.apache.spark.mllib.classification.NaiveBayes
import org.apache.spark.mllib.tree.DecisionTree
import org.apache.spark.mllib.tree.configuration.Algo
import org.apache.spark.mllib.tree.impurity.Entropy
val numIterations = 10
val maxTreeDepth = 5
```

Now, train each model in turn. First, we will train logistic regression:

```
val lrModel = LogisticRegressionWithSGD.train(data, numIterations)
```

...

14/12/06 13:41:47 INFO DAGScheduler: Job 81 finished: reduce at RDDFunctions.scala:112, took 0.011968 s

14/12/06 13:41:47 INFO GradientDescent: GradientDescent.runMiniBatchSGD finished. Last 10 stochastic losses 0.6931471805599474, 1196521.395699124, Infinity, 1861127.002201189, Infinity, 2639638.049627607, Infinity, Infinity, Infinity, Infinity

lrModel: org.apache.spark.mllib.classification.LogisticRegressionModel = (weights=[-0.11372778986947886,-0.511619752777837,

...

Next up, we will train an SVM model:

```
val svmModel = SVMWithSGD.train(data, numIterations)
```

You will see the following output:

...

14/12/06 13:43:08 INFO DAGScheduler: Job 94 finished: reduce at RDDFunctions.scala:112, took 0.007192 s

```
14/12/06 13:43:08 INFO GradientDescent: GradientDescent.runMiniBatchSGD
finished. Last 10 stochastic losses 1.0, 2398226.619666797,
2196192.9647478117, 3057987.2024311484, 271452.9038284356,
3158131.191895948, 1041799.350498323, 1507522.941537049,
1754560.9909073508, 136866.76745605646
svmModel: org.apache.spark.mllib.classification.SVMModel = (weigh
ts=[-0.12218838697834929,-0.5275107581589767,
...
```

Then, we will train the naïve Bayes model; remember to use your special non-negative feature dataset:

```
val nbModel = NaiveBayes.train(nbData)
```

The following is the output:

```
...
14/12/06 13:44:48 INFO DAGScheduler: Job 95 finished: collect at
NaiveBayes.scala:120, took 0.441273 s
nbModel: org.apache.spark.mllib.classification.NaiveBayesModel = org.
apache.spark.mllib.classification.NaiveBayesModel@666ac612
...
```

Finally, we will train our decision tree:

```
val dtModel = DecisionTree.train(data, Algo.Classification, Entropy,
maxTreeDepth)
```

The output is as follows:

```
...
14/12/06 13:46:03 INFO DAGScheduler: Job 104 finished: collectAsMap at
DecisionTree.scala:653, took 0.031338 s
...
  total: 0.343024
  findSplitsBins: 0.119499
  findBestSplits: 0.200352
  chooseSplits: 0.199705
dtModel: org.apache.spark.mllib.tree.model.DecisionTreeModel =
DecisionTreeModel classifier of depth 5 with 61 nodes
...
```

Notice that we set the mode, or `Algo`, of the decision tree to `Classification`, and we used the `Entropy` impurity measure.

Using classification models

We now have four models trained on our input labels and features. We will now see how to use these models to make predictions on our dataset. For now, we will use the same training data to illustrate the `predict` method of each model.

Generating predictions for the Kaggle/ StumbleUpon evergreen classification dataset

We will use our logistic regression model as an example (the other models are used in the same way):

```
val dataPoint = data.first
val prediction = lrModel.predict(dataPoint.features)
```

The following is the output:

prediction: Double = 1.0

We saw that for the first data point in our training dataset, the model predicted a label of 1 (that is, evergreen). Let's examine the true label for this data point:

```
val trueLabel = dataPoint.label
```

You can see the following output:

trueLabel: Double = 0.0

So, in this case, our model got it wrong!

We can also make predictions in bulk by passing in an RDD[Vector] as input:

```
val predictions = lrModel.predict(data.map(lp => lp.features))
predictions.take(5)
```

The following is the output:

Array[Double] = Array(1.0, 1.0, 1.0, 1.0, 1.0)

Evaluating the performance of classification models

When we make predictions using our model, as we did earlier, how do we know whether the predictions are good or not? We need to be able to evaluate how well our model performs. Evaluation metrics commonly used in binary classification include prediction accuracy and error, precision and recall, and area under the precision-recall curve, the **receiver operating characteristic** (**ROC**) curve, **area under ROC curve** (**AUC**), and F-measure.

Accuracy and prediction error

The prediction error for binary classification is possibly the simplest measure available. It is the number of training examples that are misclassified, divided by the total number of examples. Similarly, accuracy is the number of correctly classified examples divided by the total examples.

We can calculate the accuracy of our models in our training data by making predictions on each input feature and comparing them to the true label. We will sum up the number of correctly classified instances and divide this by the total number of data points to get the average classification accuracy:

```
val lrTotalCorrect = data.map { point =>
  if (lrModel.predict(point.features) == point.label) 1 else 0
}.sum
val lrAccuracy = lrTotalCorrect / data.count
```

The output is as follows:

`lrAccuracy: Double = 0.5146720757268425`

This gives us 51.5 percent accuracy, which doesn't look particularly impressive! Our model got only half of the training examples correct, which seems to be about as good as a random chance.

> Note that the predictions made by the model are not naturally exactly 1 or 0. The output is usually a real number that must be turned into a class prediction. This is done through use of a threshold in the classifier's decision or scoring function.
>
> For example, binary logistic regression is a probabilistic model that returns the estimated probability of class 1 in its scoring function. Thus, a decision threshold of 0.5 is typical. That is, if the estimated probability of being in class 1 is higher than 50 percent, the model decides to classify the point as class 1; otherwise, it will be classified as class 0.
>
> Note that the threshold itself is effectively a model parameter that can be tuned in some models. It also plays a role in evaluation measures, as we will see now.

What about the other models? Let's compute the accuracy for the other three:

```
val svmTotalCorrect = data.map { point =>
  if (svmModel.predict(point.features) == point.label) 1 else 0
}.sum
val nbTotalCorrect = nbData.map { point =>
  if (nbModel.predict(point.features) == point.label) 1 else 0
}.sum
```

Note that the decision tree prediction threshold needs to be specified explicitly, as highlighted here:

```
val dtTotalCorrect = data.map { point =>
  val score = dtModel.predict(point.features)
  val predicted = if (score > 0.5) 1 else 0
  if (predicted == point.label) 1 else 0
}.sum
```

We can now inspect the accuracy for the other three models.

First, the SVM model:

```
val svmAccuracy = svmTotalCorrect / numData
```

Here is the output for the SVM model:

svmAccuracy: Double = 0.5146720757268425

Next, our naïve Bayes model:

```
val nbAccuracy = nbTotalCorrect / numData
```

The output is as follows:

nbAccuracy: Double = 0.5803921568627451

Finally, we compute the accuracy for the decision tree:

```
val dtAccuracy = dtTotalCorrect / numData
```

And, the output is:

```
dtAccuracy: Double = 0.6482758620689655
```

We can see that both SVM and naïve Bayes also performed quite poorly. The decision tree model is better with 65 percent accuracy, but this is still not particularly high.

Precision and recall

In information retrieval, precision is a commonly used measure of the quality of the results, while recall is a measure of the completeness of the results.

In the binary classification context, precision is defined as the number of true positives (that is, the number of examples correctly predicted as class 1) divided by the sum of true positives and false positives (that is, the number of examples that were incorrectly predicted as class 1). Thus, we can see that a precision of 1.0 (or 100 percent) is achieved if every example predicted by the classifier to be class 1 is, in fact, in class 1 (that is, there are no false positives).

Recall is defined as the number of true positives divided by the sum of true positives and false negatives (that is, the number of examples that were in class 1, but were predicted as class 0 by the model). We can see that a recall of 1.0 (or 100 percent) is achieved if the model doesn't miss any examples that were in class 1 (that is, there are no false negatives).

Generally, precision and recall are inversely related; often, higher precision is related to lower recall and vice versa. To illustrate this, assume that we built a model that always predicted class 1. In this case, the model predictions would have no false negatives because the model always predicts 1; it will not miss any of class 1. Thus, the recall will be 1.0 for this model. On the other hand, the false positive rate could be very high, meaning precision would be low (this depends on the exact distribution of the classes in the dataset).

Precision and recall are not particularly useful as standalone metrics, but are typically used together to form an aggregate or averaged metric. Precision and recall are also dependent on the threshold selected for the model.

Intuitively, below some threshold level, a model will always predict class 1. Hence, it will have a recall of 1, but most likely, it will have low precision. At a high enough threshold, the model will always predict class 0. The model will then have a recall of 0, since it cannot achieve any true positives and will likely have many false negatives. Furthermore, its precision score will be undefined, as it will achieve zero true positives and zero false positives.

The **precision-recall** (**PR**) curve shown in the following figure plots precision against recall outcomes for a given model, as the decision threshold of the classifier is changed. The area under this PR curve is referred to as the average precision. Intuitively, an area under the PR curve of 1.0 will equate to a perfect classifier that will achieve 100 percent in both precision and recall.

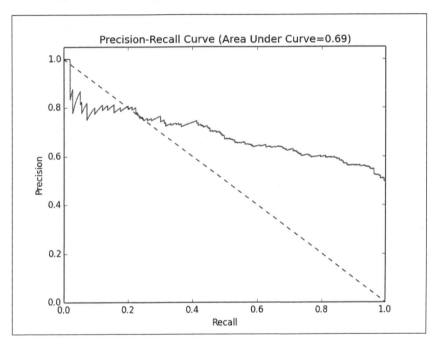

Precision-recall curve

See http://en.wikipedia.org/wiki/Precision_and_recall and http://en.wikipedia.org/wiki/Average_precision#Average_precision for more details on precision, recall, and area under the PR curve.

ROC curve and AUC

The **ROC** curve is a concept similar to the PR curve. It is a graphical illustration of the true positive rate against the false positive rate for a classifier.

The **true positive rate** (**TPR**) is the number of true positives divided by the sum of true positives and false negatives. In other words, it is the ratio of true positives to all positive examples. This is the same as the recall we saw earlier and is also commonly referred to as sensitivity.

The **false positive rate** (**FPR**) is the number of false positives divided by the sum of false positives and **true negatives** (that is, the number of examples correctly predicted as class 0). In other words, it is the ratio of false positives to all negative examples.

In a manner similar to precision and recall, the ROC curve (plotted in the following figure) represents the classifier's performance tradeoff of TPR against FPR, for different decision thresholds. Each point on the curve represents a different threshold in the decision function for the classifier.

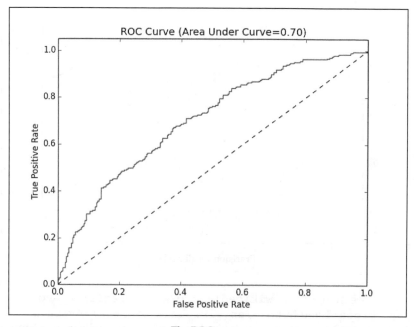

The ROC curve

The area under the ROC curve (commonly referred to as AUC) represents an average value. Again, an AUC of 1.0 will represent a perfect classifier. An area of 0.5 is referred to as the random score. Thus, a model that achieves an AUC of 0.5 is no better than randomly guessing.

As both the area under the PR curve and the area under the ROC curve are effectively normalized (with a minimum of 0 and maximum of 1), we can use these measures to compare models with differing parameter settings and even compare completely different models. Thus, these metrics are popular for model evaluation and selection purposes.

MLlib comes with a set of built-in routines to compute the area under the PR and ROC curves for binary classification. Here, we will compute these metrics for each of our models:

```
import org.apache.spark.mllib.evaluation.BinaryClassificationMetrics
val metrics = Seq(lrModel, svmModel).map { model =>
  val scoreAndLabels = data.map { point =>
    (model.predict(point.features), point.label)
  }
  val metrics = new BinaryClassificationMetrics(scoreAndLabels)
  (model.getClass.getSimpleName, metrics.areaUnderPR, metrics.areaUnderROC)
}
```

As we did previously to train the naïve Bayes model and computing accuracy, we need to use the special `nbData` version of the dataset that we created to compute the classification metrics:

```
val nbMetrics = Seq(nbModel).map{ model =>
  val scoreAndLabels = nbData.map { point =>
    val score = model.predict(point.features)
    (if (score > 0.5) 1.0 else 0.0, point.label)
  }
  val metrics = new BinaryClassificationMetrics(scoreAndLabels)
  (model.getClass.getSimpleName, metrics.areaUnderPR,
  metrics.areaUnderROC)
}
```

Note that because the `DecisionTreeModel` model does not implement the `ClassificationModel` interface that is implemented by the other three models, we need to compute the results separately for this model in the following code:

```
val dtMetrics = Seq(dtModel).map{ model =>
  val scoreAndLabels = data.map { point =>
    val score = model.predict(point.features)
    (if (score > 0.5) 1.0 else 0.0, point.label)
  }
  val metrics = new BinaryClassificationMetrics(scoreAndLabels)
```

Building a Classification Model with Spark

```
      (model.getClass.getSimpleName, metrics.areaUnderPR,
    metrics.areaUnderROC)
}
val allMetrics = metrics ++ nbMetrics ++ dtMetrics
allMetrics.foreach{ case (m, pr, roc) =>
    println(f"$m, Area under PR: ${pr * 100.0}%2.4f%%, Area under
    ROC: ${roc * 100.0}%2.4f%%")
}
```

Your output will look similar to the one here:

```
LogisticRegressionModel, Area under PR: 75.6759%, Area under ROC:
50.1418%
SVMModel, Area under PR: 75.6759%, Area under ROC: 50.1418%
NaiveBayesModel, Area under PR: 68.0851%, Area under ROC: 58.3559%
DecisionTreeModel, Area under PR: 74.3081%, Area under ROC: 64.8837%
```

We can see that all models achieve broadly similar results for the average precision metric.

Logistic regression and SVM achieve results of around 0.5 for AUC. This indicates that they do no better than random chance! Our naïve Bayes and decision tree models fare a little better, achieving an AUC of 0.58 and 0.65, respectively. Still, this is not a very good result in terms of binary classification performance.

> While we don't cover multiclass classification here, MLlib provides a similar evaluation class called `MulticlassMetrics`, which provides averaged versions of many common metrics.

Improving model performance and tuning parameters

So, what went wrong? Why have our sophisticated models achieved nothing better than random chance? Is there a problem with our models?

Recall that we started out by just throwing the data at our model. In fact, we didn't even throw all our data at the model, just the numeric columns that were easy to use. Furthermore, we didn't do a lot of analysis on these numeric features.

Feature standardization

Many models that we employ make inherent assumptions about the distribution or scale of input data. One of the most common forms of assumption is about normally-distributed features. Let's take a deeper look at the distribution of our features.

To do this, we can represent the feature vectors as a distributed matrix in MLlib, using the `RowMatrix` class. `RowMatrix` is an RDD made up of vector, where each vector is a row of our matrix.

The `RowMatrix` class comes with some useful methods to operate on the matrix, one of which is a utility to compute statistics on the columns of the matrix:

```
import org.apache.spark.mllib.linalg.distributed.RowMatrix
val vectors = data.map(lp => lp.features)
val matrix = new RowMatrix(vectors)
val matrixSummary = matrix.computeColumnSummaryStatistics()
```

The following code statement will print the mean of the matrix:

```
println(matrixSummary.mean)
```

Here is the output:

[0.41225805299526636,2.761823191986623,0.46823047328614004, ...

The following code statement will print the minimum value of the matrix:

```
println(matrixSummary.min)
```

Here is the output:

[0.0,0.0,0.0,0.0,0.0,0.0,0.0,-1.0,0.0,0.0,0.0,0.045564223,-1.0, ...

The following code statement will print the maximum value of the matrix:

```
println(matrixSummary.max)
```

The output is as follows:

[0.999426,363.0,1.0,1.0,0.980392157,0.980392157,21.0,0.25,0.0,0.444444444
, ...

The following code statement will print the variance of the matrix:

```
println(matrixSummary.variance)
```

The output of the variance is:

[0.1097424416755897,74.30082476809638,0.04126316989120246, ...

The following code statement will print the nonzero number of the matrix:

```
println(matrixSummary.numNonzeros)
```

Here is the output:

```
[5053.0,7354.0,7172.0,6821.0,6160.0,5128.0,7350.0,1257.0,0.0, ...
```

The `computeColumnSummaryStatistics` method computes a number of statistics over each column of features, including the mean and variance, storing each of these in a `Vector` with one entry per column (that is, one entry per feature in our case).

Looking at the preceding output for mean and variance, we can see quite clearly that the second feature has a much higher mean and variance than some of the other features (you will find a few other features that are similar and a few others that are more extreme). So, our data definitely does not conform to a standard Gaussian distribution in its raw form. To get the data in a more suitable form for our models, we can standardize each feature such that it has zero mean and unit standard deviation. We can do this by subtracting the column mean from each feature value and then scaling it by dividing it by the column standard deviation for the feature:

$$(x - \mu) / \sqrt{variance}$$

Practically, for each feature vector in our input dataset, we can simply perform an element-wise subtraction of the preceding mean vector from the feature vector and then perform an element-wise division of the feature vector by the vector of feature standard deviations. The standard deviation vector itself can be obtained by performing an element-wise square root operation on the variance vector.

As we mentioned in *Chapter 3, Obtaining, Processing, and Preparing Data with Spark*, we fortunately have access to a convenience method from Spark's `StandardScaler` to accomplish this.

`StandardScaler` works in much the same way as the `Normalizer` feature we used in that chapter. We will instantiate it by passing in two arguments that tell it whether to subtract the mean from the data and whether to apply standard deviation scaling. We will then fit `StandardScaler` on our input `vectors`. Finally, we will pass in an input vector to the `transform` function, which will then return a normalized vector. We will do this within the following `map` function to preserve the `label` from our dataset:

```
import org.apache.spark.mllib.feature.StandardScaler
val scaler = new StandardScaler(withMean = true, withStd =
true).fit(vectors)
val scaledData = data.map(lp => LabeledPoint(lp.label,
scaler.transform(lp.features)))
```

Our data should now be standardized. Let's inspect the first row of the original and standardized features:

```
println(data.first.features)
```

The output of the preceding line of code is as follows:

```
[0.789131,2.055555556,0.676470588,0.205882353,
```

The following code will the first row of the standardized features:

```
println(scaledData.first.features)
```

The output is as follows:

```
[1.1376439023494747,-0.08193556218743517,1.025134766284205,-0.0558631837375738,
```

As we can see, the first feature has been transformed by applying the standardization formula. We can check this by subtracting the mean (which we computed earlier) from the first feature and dividing the result by the square root of the variance (which we computed earlier):

```
println((0.789131 - 0.41225805299526636)/ math.sqrt(0.1097424416755897))
```

The result should be equal to the first element of our scaled vector:

```
1.137647336497682
```

We can now retrain our model using the standardized data. We will use only the logistic regression model to illustrate the impact of feature standardization (since the decision tree and naïve Bayes are not impacted by this):

```
val lrModelScaled = LogisticRegressionWithSGD.train(scaledData, numIterations)
val lrTotalCorrectScaled = scaledData.map { point =>
  if (lrModelScaled.predict(point.features) == point.label) 1 else
  0
}.sum
val lrAccuracyScaled = lrTotalCorrectScaled / numData
val lrPredictionsVsTrue = scaledData.map { point =>
   (lrModelScaled.predict(point.features), point.label)
}
val lrMetricsScaled = new BinaryClassificationMetrics(lrPredictionsVsTrue)
val lrPr = lrMetricsScaled.areaUnderPR
val lrRoc = lrMetricsScaled.areaUnderROC
println(f"${lrModelScaled.getClass.getSimpleName}\nAccuracy: ${lrAccuracyScaled * 100}%2.4f%%\nArea under PR: ${lrPr * 100.0}%2.4f%%\nArea under ROC: ${lrRoc * 100.0}%2.4f%%")
```

Building a Classification Model with Spark

The result should look similar to this:

```
LogisticRegressionModel
Accuracy: 62.0419%
Area under PR: 72.7254%
Area under ROC: 61.9663%
```

Simply through standardizing our features, we have improved the logistic regression performance for accuracy and AUC from 50 percent, no better than random, to 62 percent.

Additional features

We have seen that we need to be careful about standardizing and potentially normalizing our features, and the impact on model performance can be serious. In this case, we used only a portion of the features available. For example, we completely ignored the category variable and the textual content in the boilerplate variable column.

This was done for ease of illustration, but let's assess the impact of adding an additional feature such as the category feature.

First, we will inspect the categories and form a mapping of index to category, which you might recognize as the basis for a *1-of-k* encoding of this categorical feature:

```
val categories = records.map(r => r(3)).distinct.collect.zipWithIndex.toMap
val numCategories = categories.size
println(categories)
```

The output of the different categories is as follows:

```
Map("weather" -> 0, "sports" -> 6, "unknown" -> 4, "computer_internet" -> 12, "?" -> 11, "culture_politics" -> 3, "religion" -> 8, "recreation" -> 2, "arts_entertainment" -> 9, "health" -> 5, "law_crime" -> 10, "gaming" -> 13, "business" -> 1, "science_technology" -> 7)
```

The following code will print the number of categories:

```
println(numCategories)
```

Here is the output:

14

So, we will need to create a vector of length 14 to represent this feature and assign a value of 1 for the index of the relevant category for each data point. We can then prepend this new feature vector to the vector of other numerical features:

```
val dataCategories = records.map { r =>
  val trimmed = r.map(_.replaceAll("\"", ""))
  val label = trimmed(r.size - 1).toInt
  val categoryIdx = categories(r(3))
  val categoryFeatures = Array.ofDim[Double](numCategories)
  categoryFeatures(categoryIdx) = 1.0
  val otherFeatures = trimmed.slice(4, r.size - 1).map(d => if
  (d == "?") 0.0 else d.toDouble)
  val features = categoryFeatures ++ otherFeatures
  LabeledPoint(label, Vectors.dense(features))
}
println(dataCategories.first)
```

You should see output similar to what is shown here. You can see that the first part of our feature vector is now a vector of length 14 with one nonzero entry at the relevant category index:

`LabeledPoint(0.0, [0.0,1.0,0.0,0.0,0.0,0.0,0.0,0.0,0.0,0.0,0.0,0.0,0.0,0.0,0.789131,2.055555556,0.676470588,0.205882353,0.047058824,0.023529412,0.443783175,0.0,0.0,0.09077381,0.0,0.245831182,0.003883495,1.0,1.0,24.0,0.0,5424.0,170.0,8.0,0.152941176,0.079129575])`

Again, since our raw features are not standardized, we should perform this transformation using the same `StandardScaler` approach that we used earlier before training a new model on this expanded dataset:

```
val scalerCats = new StandardScaler(withMean = true, withStd = true).
fit(dataCategories.map(lp => lp.features))
val scaledDataCats = dataCategories.map(lp =>
LabeledPoint(lp.label, scalerCats.transform(lp.features)))
```

We can inspect the features before and after scaling as we did earlier:

```
println(dataCategories.first.features)
```

The output is as follows:

`0.0,1.0,0.0,0.0,0.0,0.0,0.0,0.0,0.0,0.0,0.0,0.0,0.0,0.0,0.789131,2.055555556 ...`

The following code will print the features after scaling:

```
println(scaledDataCats.first.features)
```

You will see the following on the screen:

`[-0.023261105535492967,2.720728254208072,-0.4464200056407091,-0.2205258360869135, ...`

>
> Note that while the original raw features were sparse (that is, there are many entries that are zero), if we subtract the mean from each entry, we would end up with a non-sparse (dense) representation, as can be seen in the preceding example.
>
> This is not a problem in this case as the data size is small, but often large-scale real-world problems have extremely sparse input data with many features (online advertising and text classification are good examples). In this case, it is not advisable to lose this sparsity, as the memory and processing requirements for the equivalent dense representation can quickly explode with many millions of features. We can use StandardScaler and set withMean to false to avoid this.

We're now ready to train a new logistic regression model with our expanded feature set, and then we will evaluate the performance:

```
val lrModelScaledCats = LogisticRegressionWithSGD.
train(scaledDataCats, numIterations)
val lrTotalCorrectScaledCats = scaledDataCats.map { point =>
  if (lrModelScaledCats.predict(point.features) == point.label) 1 else 0
}.sum
val lrAccuracyScaledCats = lrTotalCorrectScaledCats / numData
val lrPredictionsVsTrueCats = scaledDataCats.map { point =>
  (lrModelScaledCats.predict(point.features), point.label)
}
val lrMetricsScaledCats = new BinaryClassificationMetrics(lrPredictionsVsTrueCats)
val lrPrCats = lrMetricsScaledCats.areaUnderPR
val lrRocCats = lrMetricsScaledCats.areaUnderROC
println(f"${lrModelScaledCats.getClass.getSimpleName}\nAccuracy: ${lrAccuracyScaledCats * 100}%2.4f%%\nArea under PR: ${lrPrCats * 100.0}%2.4f%%\nArea under ROC: ${lrRocCats * 100.0}%2.4f%%")
```

You should see output similar to this one:

LogisticRegressionModel
Accuracy: 66.5720%
Area under PR: 75.7964%
Area under ROC: 66.5483%

By applying a feature standardization transformation to our data, we improved both the accuracy and AUC measures from 50 percent to 62 percent, and then, we achieved a further boost to 66 percent by adding the category feature into our model (remember to apply the standardization to our new feature set).

> Note that the best model performance in the competition was an AUC of 0.88906 (see http://www.kaggle.com/c/stumbleupon/leaderboard/private).
>
> One approach to achieving performance almost as high is outlined at http://www.kaggle.com/c/stumbleupon/forums/t/5680/beating-the-benchmark-leaderboard-auc-0-878.
>
> Notice that there are still features that we have not yet used; most notably, the text features in the boilerplate variable. The leading competition submissions predominantly use the boilerplate features and features based on the raw textual content to achieve their performance. As we saw earlier, while adding category-improved performance, it appears that most of the variables are not very useful as predictors, while the textual content turned out to be highly predictive.
>
> Going through some of the best performing approaches for these competitions can give you a good idea as to how feature extraction and engineering play a critical role in model performance.

Using the correct form of data

Another critical aspect of model performance is using the correct form of data for each model. Previously, we saw that applying a naïve Bayes model to our numerical features resulted in very poor performance. Is this because the model itself is deficient?

In this case, recall that MLlib implements a multinomial model. This model works on input in the form of non-zero count data. This can include a binary representation of categorical features (such as the *1-of-k* encoding covered previously) or frequency data (such as the frequency of occurrences of words in a document). The numerical features we used initially do not conform to this assumed input distribution, so it is probably unsurprising that the model did so poorly.

To illustrate this, we'll use only the category feature, which, when *1-of-k* encoded, is of the correct form for the model. We will create a new dataset as follows:

```
val dataNB = records.map { r =>
  val trimmed = r.map(_.replaceAll("\"", ""))
  val label = trimmed(r.size - 1).toInt
  val categoryIdx = categories(r(3))
  val categoryFeatures = Array.ofDim[Double](numCategories)
  categoryFeatures(categoryIdx) = 1.0
  LabeledPoint(label, Vectors.dense(categoryFeatures))
}
```

Next, we will train a new naïve Bayes model and evaluate its performance:

```
val nbModelCats = NaiveBayes.train(dataNB)
val nbTotalCorrectCats = dataNB.map { point =>
  if (nbModelCats.predict(point.features) == point.label) 1 else 0
}.sum
val nbAccuracyCats = nbTotalCorrectCats / numData
val nbPredictionsVsTrueCats = dataNB.map { point =>
  (nbModelCats.predict(point.features), point.label)
}
val nbMetricsCats = new BinaryClassificationMetrics(nbPredictionsVsTrueCats)
val nbPrCats = nbMetricsCats.areaUnderPR
val nbRocCats = nbMetricsCats.areaUnderROC
println(f"${nbModelCats.getClass.getSimpleName}\nAccuracy: ${nbAccuracyCats * 100}%2.4f%%\nArea under PR: ${nbPrCats * 100.0}%2.4f%%\nArea under ROC: ${nbRocCats * 100.0}%2.4f%%")
```

You should see the following output:

NaiveBayesModel
Accuracy: 60.9601%
Area under PR: 74.0522%
Area under ROC: 60.5138%

So, by ensuring that we use the correct form of input, we have improved the performance of the naïve Bayes model slightly from 58 percent to 60 percent.

Tuning model parameters

The previous section showed the impact on model performance of feature extraction and selection, as well as the form of input data and a model's assumptions around data distributions. So far, we have discussed model parameters only in passing, but they also play a significant role in model performance.

MLlib's default `train` methods use default values for the parameters of each model. Let's take a deeper look at them.

Linear models

Both logistic regression and SVM share the same parameters, because they use the same underlying optimization technique of **stochastic gradient descent** (SGD). They differ only in the loss function applied. If we take a look at the class definition for logistic regression in MLlib, we will see the following definition:

```
class LogisticRegressionWithSGD private (
   private var stepSize: Double,
   private var numIterations: Int,
   private var regParam: Double,
   private var miniBatchFraction: Double)
   extends GeneralizedLinearAlgorithm[LogisticRegressionModel] ...
```

We can see that the arguments that can be passed to the constructor are `stepSize`, `numIterations`, `regParam`, and `miniBatchFraction`. Of these, all except `regParam` are related to the underlying optimization technique.

The instantiation code for logistic regression initializes the `Gradient`, `Updater`, and `Optimizer` and sets the relevant arguments for `Optimizer` (`GradientDescent` in this case):

```
private val gradient = new LogisticGradient()
private val updater = new SimpleUpdater()
override val optimizer = new GradientDescent(gradient, updater)
   .setStepSize(stepSize)
   .setNumIterations(numIterations)
   .setRegParam(regParam)
   .setMiniBatchFraction(miniBatchFraction)
```

`LogisticGradient` sets up the logistic loss function that defines our logistic regression model.

> While a detailed treatment of optimization techniques is beyond the scope of this book, MLlib provides two optimizers for linear models: SGD and L-BFGS. L-BFGS is often more accurate and has fewer parameters to tune.
>
> SGD is the default, while L-BGFS can currently only be used directly for logistic regression via `LogisticRegressionWithLBFGS`. Try it out yourself and compare the results to those found with SGD.
>
> See http://spark.apache.org/docs/latest/mllib-optimization.html for further details.

To investigate the impact of the remaining parameter settings, we will create a helper function that will train a logistic regression model, given a set of parameter inputs. First, we will import the required classes:

```
import org.apache.spark.rdd.RDD
import org.apache.spark.mllib.optimization.Updater
import org.apache.spark.mllib.optimization.SimpleUpdater
import org.apache.spark.mllib.optimization.L1Updater
import org.apache.spark.mllib.optimization.SquaredL2Updater
import org.apache.spark.mllib.classification.ClassificationModel
```

Next, we will define our helper function to train a mode given a set of inputs:

```
def trainWithParams(input: RDD[LabeledPoint], regParam: Double,
numIterations: Int, updater: Updater, stepSize: Double) = {
  val lr = new LogisticRegressionWithSGD
  lr.optimizer.setNumIterations(numIterations).
  setUpdater(updater).setRegParam(regParam).setStepSize(stepSize)
  lr.run(input)
}
```

Finally, we will create a second helper function to take the input data and a classification model and generate the relevant AUC metrics:

```
def createMetrics(label: String, data: RDD[LabeledPoint], model:
ClassificationModel) = {
  val scoreAndLabels = data.map { point =>
    (model.predict(point.features), point.label)
  }
  val metrics = new BinaryClassificationMetrics(scoreAndLabels)
  (label, metrics.areaUnderROC)
}
```

We will also cache our scaled dataset, including categories, to speed up the multiple model training runs that we will be using to explore these different parameter settings:

```
scaledDataCats.cache
```

Iterations

Many machine learning methods are iterative in nature, converging to a solution (the optimal weight vector that minimizes the chosen loss function) over a number of iteration steps. SGD typically requires relatively few iterations to converge to a reasonable solution but can be run for more iterations to improve the solution. We can see this by trying a few different settings for the numIterations parameter and comparing the AUC results:

```
val iterResults = Seq(1, 5, 10, 50).map { param =>
  val model = trainWithParams(scaledDataCats, 0.0, param, new SimpleUpdater, 1.0)
  createMetrics(s"$param iterations", scaledDataCats, model)
}
iterResults.foreach { case (param, auc) => println(f"$param, AUC = ${auc * 100}%2.2f%%") }
```

Your output should look like this:

```
1 iterations, AUC = 64.97%
5 iterations, AUC = 66.62%
10 iterations, AUC = 66.55%
50 iterations, AUC = 66.81%
```

So, we can see that the number of iterations has minor impact on the results once a certain number of iterations have been completed.

Step size

In SGD, the step size parameter controls how far in the direction of the steepest gradient the algorithm takes a step when updating the model weight vector after each training example. A larger step size might speed up convergence, but a step size that is too large might cause problems with convergence as good solutions are overshot.

We can see the impact of changing the step size here:

```
val stepResults = Seq(0.001, 0.01, 0.1, 1.0, 10.0).map { param =>
  val model = trainWithParams(scaledDataCats, 0.0, numIterations, new SimpleUpdater, param)
  createMetrics(s"$param step size", scaledDataCats, model)
}
stepResults.foreach { case (param, auc) => println(f"$param, AUC = ${auc * 100}%2.2f%%") }
```

This will give us the following results, which show that increasing the step size too much can begin to negatively impact performance.

```
0.001 step size, AUC = 64.95%
0.01 step size, AUC = 65.00%
0.1 step size, AUC = 65.52%
1.0 step size, AUC = 66.55%
10.0 step size, AUC = 61.92%
```

Regularization

We briefly touched on the `Updater` class in the preceding logistic regression code. An `Updater` class in MLlib implements regularization. Regularization can help avoid over-fitting of a model to training data by effectively penalizing model complexity. This can be done by adding a term to the loss function that acts to increase the loss as a function of the model weight vector.

Regularization is almost always required in real use cases, but is of particular importance when the feature dimension is very high (that is, the effective number of variable weights that can be learned is high) relative to the number of training examples.

When regularization is absent or low, models can tend to over-fit. Without regularization, most models will over-fit on a training dataset. This is a key reason behind the use of cross-validation techniques for model fitting (which we will cover now).

Conversely, since applying regularization encourages simpler models, model performance can suffer when regularization is high through under-fitting the data.

The forms of regularization available in MLlib are:

- `SimpleUpdater`: This equates to no regularization and is the default for logistic regression
- `SquaredL2Updater`: This implements a regularizer based on the squared L2-norm of the weight vector; this is the default for SVM models
- `L1Updater`: This applies a regularizer based on the L1-norm of the weight vector; this can lead to sparse solutions in the weight vector (as less important weights are pulled towards zero)

> Regularization and its relation to optimization is a broad and heavily researched area. Some more information is available from the following links:
> - General regularization overview: http://en.wikipedia.org/wiki/Regularization_(mathematics)
> - L2 regularization: http://en.wikipedia.org/wiki/Tikhonov_regularization
> - Over-fitting and under-fitting: http://en.wikipedia.org/wiki/Overfitting
> - Detailed overview of over-fitting and L1 versus L2 regularization: http://citeseerx.ist.psu.edu/viewdoc/download?doi=10.1.1.92.9860&rep=rep1&type=pdf

Let's explore the impact of a range of regularization parameters using `SquaredL2Updater`:

```
val regResults = Seq(0.001, 0.01, 0.1, 1.0, 10.0).map { param =>
  val model = trainWithParams(scaledDataCats, param, numIterations, new SquaredL2Updater, 1.0)
  createMetrics(s"$param L2 regularization parameter", scaledDataCats, model)
}
regResults.foreach { case (param, auc) => println(f"$param, AUC = ${auc * 100}%2.2f%%") }
```

Your output should look like this:

```
0.001 L2 regularization parameter, AUC = 66.55%
0.01 L2 regularization parameter, AUC = 66.55%
0.1 L2 regularization parameter, AUC = 66.63%
1.0 L2 regularization parameter, AUC = 66.04%
10.0 L2 regularization parameter, AUC = 35.33%
```

As we can see, at low levels of regularization, there is not much impact in model performance. However, as we increase regularization, we can see the impact of under-fitting on our model evaluation.

> You will find similar results when using the L1 regularization. Give it a try by performing the same evaluation of regularization parameter against the AUC measure for `L1Updater`.

Decision trees

The decision tree model we trained earlier was the best performer on the raw data that we first used. We set a parameter called `maxDepth`, which controls the maximum depth of the tree and, thus, the complexity of the model. Deeper trees result in more complex models that will be able to fit the data better.

For classification problems, we can also select between two measures of impurity: `Gini` and `Entropy`.

Tuning tree depth and impurity

We will illustrate the impact of tree depth in a similar manner as we did for our logistic regression model.

First, we will need to create another helper function in the Spark shell:

```
import org.apache.spark.mllib.tree.impurity.Impurity
import org.apache.spark.mllib.tree.impurity.Entropy
import org.apache.spark.mllib.tree.impurity.Gini

def trainDTWithParams(input: RDD[LabeledPoint], maxDepth: Int,
impurity: Impurity) = {
  DecisionTree.train(input, Algo.Classification, impurity,
maxDepth)
}
```

Now, we're ready to compute our AUC metric for different settings of tree depth. We will simply use our original dataset in this example since we do not need the data to be standardized.

> Note that decision tree models generally do not require features to be standardized or normalized, nor do they require categorical features to be binary-encoded.

First, train the model using the `Entropy` impurity measure and varying tree depths:

```
val dtResultsEntropy = Seq(1, 2, 3, 4, 5, 10, 20).map { param =>
  val model = trainDTWithParams(data, param, Entropy)
  val scoreAndLabels = data.map { point =>
    val score = model.predict(point.features)
    (if (score > 0.5) 1.0 else 0.0, point.label)
  }
```

```
    val metrics = new BinaryClassificationMetrics(scoreAndLabels)
    (s"$param tree depth", metrics.areaUnderROC)
}
dtResultsEntropy.foreach { case (param, auc) => println(f"$param, 
AUC = ${auc * 100}%2.2f%%") }
```

This should output the results shown here:

```
1 tree depth, AUC = 59.33%
2 tree depth, AUC = 61.68%
3 tree depth, AUC = 62.61%
4 tree depth, AUC = 63.63%
5 tree depth, AUC = 64.88%
10 tree depth, AUC = 76.26%
20 tree depth, AUC = 98.45%
```

Next, we will perform the same computation using the `Gini` impurity measure (we omitted the code as it is very similar, but it can be found in the code bundle). Your results should look something like this:

```
1 tree depth, AUC = 59.33%
2 tree depth, AUC = 61.68%
3 tree depth, AUC = 62.61%
4 tree depth, AUC = 63.63%
5 tree depth, AUC = 64.89%
10 tree depth, AUC = 78.37%
20 tree depth, AUC = 98.87%
```

As you can see from the preceding results, increasing the tree depth parameter results in a more accurate model (as expected since the model is allowed to get more complex with greater tree depth). It is very likely that at higher tree depths, the model will over-fit the dataset significantly.

There is very little difference in performance between the two impurity measures.

The naïve Bayes model

Finally, let's see the impact of changing the `lambda` parameter for naïve Bayes. This parameter controls additive smoothing, which handles the case when a class and feature value do not occur together in the dataset.

See http://en.wikipedia.org/wiki/Additive_smoothing for more details on additive smoothing.

We will take the same approach as we did earlier, first creating a convenience training function and training the model with varying levels of `lambda`:

```
def trainNBWithParams(input: RDD[LabeledPoint], lambda: Double) = {
  val nb = new NaiveBayes
  nb.setLambda(lambda)
  nb.run(input)
}
val nbResults = Seq(0.001, 0.01, 0.1, 1.0, 10.0).map { param =>
  val model = trainNBWithParams(dataNB, param)
  val scoreAndLabels = dataNB.map { point =>
    (model.predict(point.features), point.label)
  }
  val metrics = new BinaryClassificationMetrics(scoreAndLabels)
  (s"$param lambda", metrics.areaUnderROC)
}
nbResults.foreach { case (param, auc) => println(f"$param, AUC = ${auc * 100}%2.2f%%")
}
```

The results of the training are as follows:

```
0.001 lambda, AUC = 60.51%
0.01 lambda, AUC = 60.51%
0.1 lambda, AUC = 60.51%
1.0 lambda, AUC = 60.51%
10.0 lambda, AUC = 60.51%
```

We can see that `lambda` has no impact in this case, since it will not be a problem if the combination of feature and class label not occurring together in the dataset.

Cross-validation

So far in this book, we have only briefly mentioned the idea of cross-validation and out-of-sample testing. Cross-validation is a critical part of real-world machine learning and is central to many model selection and parameter tuning pipelines.

The general idea behind cross-validation is that we want to know how our model will perform on unseen data. Evaluating this on real, live data (for example, in a production system) is risky, because we don't really know whether the trained model is the best in the sense of being able to make accurate predictions on new data. As we saw previously with regard to regularization, our model might have over-fit the training data and be poor at making predictions on data it has not been trained on.

Cross-validation provides a mechanism where we use part of our available dataset to train our model and another part to evaluate the performance of this model. As the model is tested on data that it has not seen during the training phase, its performance, when evaluated on this part of the dataset, gives us an estimate as to how well our model generalizes for the new data points.

Here, we will implement a simple cross-validation evaluation approach using a train-test split. We will divide our dataset into two non-overlapping parts. The first dataset is used to train our model and is called the training set. The second dataset, called the test set or hold-out set, is used to evaluate the performance of our model using our chosen evaluation measure. Common splits used in practice include 50/50, 60/40, and 80/20 splits, but you can use any split as long as the training set is not too small for the model to learn (generally, at least 50 percent is a practical minimum).

In many cases, three sets are created: a training set, an evaluation set (which is used like the above test set to tune the model parameters such as lambda and step size), and a test set (which is never used to train a model or tune any parameters, but is only used to generate an estimated true performance on completely unseen data).

Here, we will explore a simple train-test split approach. There are many cross-validation techniques that are more exhaustive and complex.

One popular example is K-fold cross-validation, where the dataset is split into K non-overlapping folds. The model is trained on K-1 folds of data and tested on the remaining, held-out fold. This is repeated K times, and the results are averaged to give the cross-validation score. The train-test split is effectively like two-fold cross-validation.

Other approaches include leave-one-out cross-validation and random sampling. See the article at http://en.wikipedia.org/wiki/Cross-validation_(statistics) for further details.

First, we will split our dataset into a 60 percent training set and a 40 percent test set (we will use a constant random seed of 123 here to ensure that we get the same results for ease of illustration):

```
val trainTestSplit = scaledDataCats.randomSplit(Array(0.6, 0.4), 123)
val train = trainTestSplit(0)
val test = trainTestSplit(1)
```

Next, we will compute the evaluation metric of interest (again, we will use AUC) for a range of regularization parameter settings. Note that here we will use a finer-grained step size between the evaluated regularization parameters to better illustrate the differences in AUC, which are very small in this case:

```
val regResultsTest = Seq(0.0, 0.001, 0.0025, 0.005, 0.01).map { param =>
   val model = trainWithParams(train, param, numIterations, new SquaredL2Updater, 1.0)
    createMetrics(s"$param L2 regularization parameter", test, model)
}
regResultsTest.foreach { case (param, auc) => println(f"$param, AUC = ${auc * 100}%2.6f%%")
}
```

This will compute the results of training on the training set and the results of evaluating on the test set, as shown here:

```
0.0 L2 regularization parameter, AUC = 66.480874%
0.001 L2 regularization parameter, AUC = 66.480874%
0.0025 L2 regularization parameter, AUC = 66.515027%
0.005 L2 regularization parameter, AUC = 66.515027%
0.01 L2 regularization parameter, AUC = 66.549180%
```

Now, let's compare this to the results of training and testing on the training set (this is what we were doing previously by training and testing on all data). Again, we will omit the code as it is very similar (but it is available in the code bundle):

```
0.0 L2 regularization parameter, AUC = 66.260311%
0.001 L2 regularization parameter, AUC = 66.260311%
0.0025 L2 regularization parameter, AUC = 66.260311%
0.005 L2 regularization parameter, AUC = 66.238294%
0.01 L2 regularization parameter, AUC = 66.238294%
```

So, we can see that when we train and evaluate our model on the same dataset, we generally achieve the highest performance when regularization is lower. This is because our model has seen all the data points, and with low levels of regularization, it can over-fit the data set and achieve higher performance.

In contrast, when we train on one dataset and test on another, we see that generally a slightly higher level of regularization results in better test set performance.

In cross-validation, we would typically find the parameter settings (including regularization as well as the various other parameters such as step size and so on) that result in the best test set performance. We would then use these parameter settings to retrain the model on all of our data in order to use it to make predictions on new data.

> Recall from *Chapter 4, Building a Recommendation Engine with Spark*, that we did not cover cross-validation. You can apply the same techniques we used earlier to split the ratings dataset from that chapter into a training and test dataset. You can then try out different parameter settings on the training set while evaluating the MSE and MAP performance metrics on the test set in a manner similar to what we did earlier. Give it a try!

Summary

In this chapter, we covered the various classification models available in Spark MLlib, and we saw how to train models on input data and how to evaluate their performance using standard metrics and measures. We also explored how to apply some of the techniques previously introduced to transform our features. Finally, we investigated the impact of using the correct input data format or distribution on model performance, and we also saw the impact of adding more data to our model, tuning model parameters, and implementing cross-validation.

In the next chapter, we will take a similar approach to delve into MLlib's regression models.

Building a Regression Model with Spark

In this chapter, we will build on what we covered in *Chapter 5, Building a Classification Model with Spark*. While classification models deal with outcomes that represent discrete classes, regression models are concerned with target variables that can take any real value. The underlying principle is very similar—we wish to find a model that maps input features to predicted target variables. Like classification, regression is also a form of supervised learning.

Regression models can be used to predict just about any variable of interest. A few examples include the following:

- Predicting stock returns and other economic variables
- Predicting loss amounts for loan defaults (this can be combined with a classification model that predicts the probability of default, while the regression model predicts the amount in the case of a default)
- Recommendations (the Alternating Least Squares factorization model from *Chapter 4, Building a Recommendation Engine with Spark,* uses linear regression in each iteration)
- Predicting **customer lifetime value** (**CLTV**) in a retail, mobile, or other business, based on user behavior and spending patterns

In the following sections, we will:

- Introduce the various types of regression models available in MLlib
- Explore feature extraction and target variable transformation for regression models

- Train a number of regression models using MLlib
- See how to make predictions using the trained models
- Investigate the impact on performance of various parameter settings for regression using cross-validation

Types of regression models

Spark's MLlib library offers two broad classes of regression models: linear models and decision tree regression models.

Linear models are essentially the same as their classification counterparts, the only difference is that linear regression models use a different loss function, related link function, and decision function. MLlib provides a standard least squares regression model (although other types of generalized linear models for regression are planned).

Decision trees can also be used for regression by changing the impurity measure.

Least squares regression

You might recall from *Chapter 5, Building a Classification Model with Spark*, that there are a variety of loss functions that can be applied to generalized linear models. The loss function used for least squares is the squared loss, which is defined as follows:

$$\frac{1}{2} (w^T x - y)^2$$

Here, as for the classification setting, y is the target variable (this time, real valued), w is the weight vector, and x is the feature vector.

The related link function is the identity link, and the decision function is also the identity function, as generally, no thresholding is applied in regression. So, the model's prediction is simply $y = w^T x$.

The standard least squares regression in MLlib does not use regularization. Looking at the squared loss function, we can see that the loss applied to incorrectly predicted points will be magnified since the loss is squared. This means that least squares regression is susceptible to outliers in the dataset and also to over-fitting. Generally, as for classification, we should apply some level of regularization in practice.

Linear regression with L2 regularization is commonly referred to as ridge regression, while applying L1 regularization is called the **lasso**.

> See the section on linear least squares in the Spark MLlib documentation at `http://spark.apache.org/docs/latest/mllib-linear-methods.html#linear-least-squares-lasso-and-ridge-regression` for further information.

Decision trees for regression

Just like using linear models for regression tasks involves changing the loss function used, using decision trees for regression involves changing the measure of the node impurity used. The impurity metric is called **variance** and is defined in the same way as the squared loss for least squares linear regression.

> See the *MLlib - Decision Tree* section in the Spark documentation at `http://spark.apache.org/docs/latest/mllib-decision-tree.html` for further details on the decision tree algorithm and impurity measure for regression.

Now, we will plot a simple example of a regression problem with only one input variable shown on the *x* axis and the target variable on the *y* axis. The linear model prediction function is shown by a red dashed line, while the decision tree prediction function is shown by a green dashed line. We can see that the decision tree allows a more complex, nonlinear model to be fitted to the data.

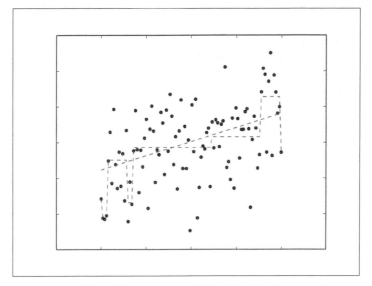

Linear model and decision tree prediction functions for regression

Extracting the right features from your data

As the underlying models for regression are the same as those for the classification case, we can use the same approach to create input features. The only practical difference is that the target is now a real-valued variable, as opposed to a categorical one. The LabeledPoint class in MLlib already takes this into account, as the label field is of the Double type, so it can handle both cases.

Extracting features from the bike sharing dataset

To illustrate the concepts in this chapter, we will be using the bike sharing dataset. This dataset contains hourly records of the number of bicycle rentals in the capital bike sharing system. It also contains variables related to date and time, weather, and seasonal and holiday information.

> The dataset is available at http://archive.ics.uci.edu/ml/datasets/Bike+Sharing+Dataset.
>
> Click on the **Data Folder** link and then download the Bike-Sharing-Dataset.zip file.
>
> The bike sharing data was enriched with weather and seasonal data by Hadi Fanaee-T at the University of Porto and used in the following paper:
>
> Fanaee-T, Hadi and Gama Joao, Event labeling combining ensemble detectors and background knowledge, *Progress in Artificial Intelligence*, pp. 1-15, Springer Berlin Heidelberg, 2013.
>
> The paper is available at http://link.springer.com/article/10.1007%2Fs13748-013-0040-3.

Once you have downloaded the Bike-Sharing-Dataset.zip file, unzip it. This will create a directory called Bike-Sharing-Dataset, which contains the day.csv, hour.csv, and the Readme.txt files.

The Readme.txt file contains information on the dataset, including the variable names and descriptions. Take a look at the file, and you will see that we have the following variables available:

- instant: This is the record ID
- dteday: This is the raw date

- `season`: This is different seasons such as spring, summer, winter, and fall
- `yr`: This is the year (2011 or 2012)
- `mnth`: This is the month of the year
- `hr`: This is the hour of the day
- `holiday`: This is whether the day was a holiday or not
- `weekday`: This is the day of the week
- `workingday`: This is whether the day was a working day or not
- `weathersit`: This is a categorical variable that describes the weather at a particular time
- `temp`: This is the normalized temperature
- `atemp`: This is the normalized apparent temperature
- `hum`: This is the normalized humidity
- `windspeed`: This is the normalized wind speed
- `cnt`: This is the target variable, that is, the count of bike rentals for that hour

We will work with the hourly data contained in `hour.csv`. If you look at the first line of the dataset, you will see that it contains the column names as a header. You can do this by running the following command:

```
>head -1 hour.csv
```

This should output the following result:

```
instant,dteday,season,yr,mnth,hr,holiday,weekday,workingday,weathersit,temp,atemp,hum,windspeed,casual,registered,cnt
```

Before we work with the data in Spark, we will again remove the header from the first line of the file using the same `sed` command that we used previously to create a new file called `hour_noheader.csv`:

```
>sed 1d hour.csv > hour_noheader.csv
```

Since we will be doing some plotting of our dataset later on, we will use the Python shell for this chapter. This also serves to illustrate how to use MLlib's linear model and decision tree functionality from PySpark.

Start up your PySpark shell from your Spark installation directory. If you want to use IPython, which we highly recommend, remember to include the `IPYTHON=1` environment variable together with the `pylab` functionality:

```
>IPYTHON=1 IPYTHON_OPTS="–pylab" ./bin/pyspark
```

If you prefer to use IPython Notebook, you can start it with the following command:

`>IPYTHON=1 IPYTHON_OPTS=notebook ./bin/pyspark`

You can type all the code that follows for the remainder of this chapter directly into your PySpark shell (or into IPython Notebook if you wish to use it).

> Recall that we used the IPython shell in *Chapter 3, Obtaining, Processing, and Preparing Data with Spark*. Take a look at that chapter and the code bundle for instructions to install IPython.

We'll start as usual by loading the dataset and inspecting it:

```
path = "/PATH/hour_noheader.csv"
raw_data = sc.textFile(path)
num_data = raw_data.count()
records = raw_data.map(lambda x: x.split(","))
first = records.first()
print first
print num_data
```

You should see the following output:

```
[u'1', u'2011-01-01', u'1', u'0', u'1', u'0', u'0', u'6', u'0', u'1',
u'0.24', u'0.2879', u'0.81', u'0', u'3', u'13', u'16']
17379
```

So, we have `17,379` hourly records in our dataset. We have inspected the column names already. We will ignore the record ID and raw date columns. We will also ignore the `casual` and `registered` count target variables and focus on the overall count variable, `cnt` (which is the sum of the other two counts). We are left with 12 variables. The first eight are categorical, while the last 4 are normalized real-valued variables.

To deal with the eight categorical variables, we will use the binary encoding approach with which you should be quite familiar by now. The four real-valued variables will be left as is.

We will first cache our dataset, since we will be reading from it many times:

```
records.cache()
```

In order to extract each categorical feature into a binary vector form, we will need to know the feature mapping of each feature value to the index of the nonzero value in our binary vector. Let's define a function that will extract this mapping from our dataset for a given column:

```
def get_mapping(rdd, idx):
    return rdd.map(lambda fields: fields[idx]).distinct().zipWithIndex().collectAsMap()
```

Our function first maps the field to its unique values and then uses the `zipWithIndex` transformation to zip the value up with a unique index such that a key-value RDD is formed, where the key is the variable and the value is the index. This index will be the index of the nonzero entry in the binary vector representation of the feature. We will finally collect this RDD back to the driver as a Python dictionary.

We can test our function on the third variable column (index 2):

```
print "Mapping of first categorical feasture column: %s" % get_mapping(records, 2)
```

The preceding line of code will give us the following output:

Mapping of first categorical feasture column: {u'1': 0, u'3': 2, u'2': 1, u'4': 3}

Now, we can apply this function to each categorical column (that is, for variable indices 2 to 9):

```
mappings = [get_mapping(records, i) for i in range(2,10)]
cat_len = sum(map(len, mappings))
num_len = len(records.first()[11:15])
total_len = num_len + cat_len
```

We now have the mappings for each variable, and we can see how many values in total we need for our binary vector representation:

```
print "Feature vector length for categorical features: %d" % cat_len
print "Feature vector length for numerical features: %d" % num_len
print "Total feature vector length: %d" % total_len
```

The output of the preceding code is as follows:

Feature vector length for categorical features: 57

Feature vector length for numerical features: 4

Total feature vector length: 61

Creating feature vectors for the linear model

The next step is to use our extracted mappings to convert the categorical features to binary-encoded features. Again, it will be helpful to create a function that we can apply to each record in our dataset for this purpose. We will also create a function to extract the target variable from each record. We will need to import `numpy` for linear algebra utilities and MLlib's `LabeledPoint` class to wrap our feature vectors and target variables:

```
from pyspark.mllib.regression import LabeledPoint
import numpy as np

def extract_features(record):
  cat_vec = np.zeros(cat_len)
  i = 0
  step = 0
  for field in record[2:9]:
    m = mappings[i]
    idx = m[field]
    cat_vec[idx + step] = 1
    i = i + 1
    step = step + len(m)
  num_vec = np.array([float(field) for field in record[10:14]])
  return np.concatenate((cat_vec, num_vec))

def extract_label(record):
  return float(record[-1])
```

In the preceding `extract_features` function, we ran through each column in the row of data. We extracted the binary encoding for each variable in turn from the mappings we created previously. The `step` variable ensures that the nonzero feature index in the full feature vector is correct (and is somewhat more efficient than, say, creating many smaller binary vectors and concatenating them). The numeric vector is created directly by first converting the data to floating point numbers and wrapping these in a `numpy` array. The resulting two vectors are then concatenated. The `extract_label` function simply converts the last column variable (the count) into a float.

With our utility functions defined, we can proceed with extracting feature vectors and labels from our data records:

```
data = records.map(lambda r: LabeledPoint(extract_label(r), extract_features(r)))
```

Let's inspect the first record in the extracted feature RDD:

```
first_point = data.first()
print "Raw data: " + str(first[2:])
print "Label: " + str(first_point.label)
print "Linear Model feature vector:\n" + str(first_point.features)
print "Linear Model feature vector length: " + str(len(first_point.features))
```

You should see output similar to the following:

```
Raw data: [u'1', u'0', u'1', u'0', u'0', u'6', u'0', u'1', u'0.24',
u'0.2879', u'0.81', u'0', u'3', u'13', u'16']
Label: 16.0
Linear Model feature vector: [1.0,0.0,0.0,0.0,0.0,1.0,0.0,1.0,0.0,0.0,0.0
,0.0,0.0,0.0,0.0,0.0,0.0,0.0,0.0,0.0,0.0,0.0,0.0,0.0,0.0,0.0,0.0,0.0,0.0,
0.0,0.0,0.0,0.0,0.0,0.0,1.0,0.0,0.0,0.0,0.0,0.0,0.0,0.0,1.0,0.0,0.0,0.0,0
.0,0.0,0.0,1.0,0.0,1.0,0.0,0.0,0.0,0.0,0.24,0.2879,0.81,0.0]
Linear Model feature vector length: 61
```

As we can see, we converted the raw data into a feature vector made up of the binary categorical and real numeric features, and we indeed have a total vector length of 61.

Creating feature vectors for the decision tree

As we have seen, decision tree models typically work on raw features (that is, it is not required to convert categorical features into a binary vector encoding; they can, instead, be used directly). Therefore, we will create a separate function to extract the decision tree feature vector, which simply converts all the values to floats and wraps them in a numpy array:

```
def extract_features_dt(record):
    return np.array(map(float, record[2:14]))
data_dt = records.map(lambda r: LabeledPoint(extract_label(r),
extract_features_dt(r)))
first_point_dt = data_dt.first()
print "Decision Tree feature vector: " + str(first_point_dt.features)
print "Decision Tree feature vector length: " +
str(len(first_point_dt.features))
```

The following output shows the extracted feature vector, and we can see that we have a vector length of 12, which matches the number of raw variables we are using:

```
Decision Tree feature vector: [1.0,0.0,1.0,0.0,0.0,6.0,0.0,1.0,0.24,0.287
9,0.81,0.0]
Decision Tree feature vector length: 12
```

Training and using regression models

Training for regression models using decision trees and linear models follows the same procedure as for classification models. We simply pass the training data contained in a `[LabeledPoint]` RDD to the relevant `train` method. Note that in Scala, if we wanted to customize the various model parameters (such as regularization and step size for the SGD optimizer), we are required to instantiate a new model instance and use the `optimizer` field to access these available parameter setters.

In Python, we are provided with a convenience method that gives us access to all the available model arguments, so we only have to use this one entry point for training. We can see the details of these convenience functions by importing the relevant modules and then calling the `help` function on the `train` methods:

```
from pyspark.mllib.regression import LinearRegressionWithSGD
from pyspark.mllib.tree import DecisionTree
help(LinearRegressionWithSGD.train)
```

Doing this for the linear model outputs the following documentation:

```
Help on method train in module pyspark.mllib.regression:

train(cls, data, iterations=100, step=1.0, miniBatchFraction=1.0, initialWeights=None, regParam=0.0, regType=None, intercept=Fa
lse) method of __builtin__.type instance
    Train a linear regression model on the given data.

    :param data:           The training data.
    :param iterations:     The number of iterations (default: 100).
    :param step:           The step parameter used in SGD
                           (default: 1.0).
    :param miniBatchFraction: Fraction of data to be used for each SGD
                           iteration.
    :param initialWeights: The initial weights (default: None).
    :param regParam:       The regularizer parameter (default: 0.0).
    :param regType:        The type of regularizer used for training
                           our model.

                           :Allowed values:
                              - "l1" for using L1 regularization (lasso),
                              - "l2" for using L2 regularization (ridge),
                              - None for no regularization

                           (default: None)

    @param intercept:      Boolean parameter which indicates the use
                           or not of the augmented representation for
                           training data (i.e. whether bias features
                           are activated or not).
```

Linear regression help documentation

We can see from the linear regression documentation that we need to pass in the training data at a minimum, but we can set any of the other model parameters using this `train` method.

Similarly, for the decision tree model, which has a `trainRegressor` method (in addition to a `trainClassifier` method for classification models):

```
help(DecisionTree.trainRegressor)
```

The preceding code would display the following documentation:

```
Help on method trainRegressor in module pyspark.mllib.tree:

trainRegressor(cls, data, categoricalFeaturesInfo, impurity='variance', maxDepth=5, maxBins=32, minInstancesPerNode=1, minInfoG
ain=0.0) method of __builtin__.type instance
    Train a DecisionTreeModel for regression.

    :param data: Training data: RDD of LabeledPoint.
                 Labels are real numbers.
    :param categoricalFeaturesInfo: Map from categorical feature index
                                    to number of categories.
                                    Any feature not in this map
                                    is treated as continuous.
    :param impurity: Supported values: "variance"
    :param maxDepth: Max depth of tree.
                     E.g., depth 0 means 1 leaf node.
                     Depth 1 means 1 internal node + 2 leaf nodes.
    :param maxBins: Number of bins used for finding splits at each node.
    :param minInstancesPerNode: Min number of instances required at child
                                nodes to create the parent split
    :param minInfoGain: Min info gain required to create a split
    :return: DecisionTreeModel

    Example usage:

    >>> from pyspark.mllib.regression import LabeledPoint
    >>> from pyspark.mllib.tree import DecisionTree
    >>> from pyspark.mllib.linalg import SparseVector
    >>>
    >>> sparse_data = [
    ...     LabeledPoint(0.0, SparseVector(2, {0: 0.0})),
    ...     LabeledPoint(1.0, SparseVector(2, {1: 1.0})),
    ...     LabeledPoint(0.0, SparseVector(2, {0: 0.0})),
    ...     LabeledPoint(1.0, SparseVector(2, {1: 2.0}))
    ... ]
    >>>
    >>> model = DecisionTree.trainRegressor(sc.parallelize(sparse_data), {})
    >>> model.predict(SparseVector(2, {1: 1.0}))
    1.0
    >>> model.predict(SparseVector(2, {1: 0.0}))
    0.0
    >>> rdd = sc.parallelize([[0.0, 1.0], [0.0, 0.0]])
    >>> model.predict(rdd).collect()
    [1.0, 0.0]
```

Decision tree regression help documentation

Training a regression model on the bike sharing dataset

We're ready to use the features we have extracted to train our models on the bike sharing data. First, we'll train the linear regression model and take a look at the first few predictions that the model makes on the data:

```
linear_model = LinearRegressionWithSGD.train(data, iterations=10, step=0.1, intercept=False)
```

```
true_vs_predicted = data.map(lambda p: (p.label, linear_model.
predict(p.features)))
print "Linear Model predictions: " +
str(true_vs_predicted.take(5))
```

Note that we have not used the default settings for iterations and step here. We've changed the number of iterations so that the model does not take too long to train. As for the step size, you will see why this has been changed from the default a little later. You will see the following output:

Linear Model predictions: [(16.0, 119.30920003093595), (40.0, 116.95463511937379), (32.0, 116.57294610647752), (13.0, 116.43535423855654), (1.0, 116.221247828503)]

Next, we will train the decision tree model simply using the default arguments to the trainRegressor method (which equates to using a tree depth of 5). Note that we need to pass in the other form of the dataset, data_dt, that we created from the raw feature values (as opposed to the binary encoded features that we used for the preceding linear model).

We also need to pass in an argument for categoricalFeaturesInfo. This is a dictionary that maps the categorical feature index to the number of categories for the feature. If a feature is not in this mapping, it will be treated as continuous. For our purposes, we will leave this as is, passing in an empty mapping:

```
dt_model = DecisionTree.trainRegressor(data_dt,{})
preds = dt_model.predict(data_dt.map(lambda p: p.features))
actual = data.map(lambda p: p.label)
true_vs_predicted_dt = actual.zip(preds)
print "Decision Tree predictions: " + str(true_vs_predicted_
dt.take(5))
print "Decision Tree depth: " + str(dt_model.depth())
print "Decision Tree number of nodes: " + str(dt_model.numNodes())
```

This should output these predictions:

Decision Tree predictions: [(16.0, 54.913223140495866), (40.0, 54.913223140495866), (32.0, 53.171052631578945), (13.0, 14.284023668639053), (1.0, 14.284023668639053)]

Decision Tree depth: 5

Decision Tree number of nodes: 63

> This is not as bad as it sounds. While we do not cover it here, the Python code included with this chapter's code bundle includes an example of using `categoricalFeaturesInfo`. It does not make a large difference to performance in this case.

From a quick glance at these predictions, it appears that the decision tree might do better, as the linear model is quite a way off in its predictions. However, we will apply more stringent evaluation methods to find out.

Evaluating the performance of regression models

We saw in *Chapter 5, Building a Classification Model with Spark*, that evaluation methods for classification models typically focus on measurements related to predicted class memberships relative to the actual class memberships. These are binary outcomes (either the predicted class is correct or incorrect), and it is less important whether the model just barely predicted correctly or not; what we care most about is the number of correct and incorrect predictions.

When dealing with regression models, it is very unlikely that our model will precisely predict the target variable, because the target variable can take on any real value. However, we would naturally like to understand how far away our predicted values are from the true values, so will we utilize a metric that takes into account the overall deviation.

Some of the standard evaluation metrics used to measure the performance of regression models include the **Mean Squared Error** (**MSE**) and **Root Mean Squared Error** (**RMSE**), the **Mean Absolute Error** (**MAE**), the R-squared coefficient, and many others.

Mean Squared Error and Root Mean Squared Error

MSE is the average of the squared error that is used as the loss function for least squares regression:

$$\sum_{i=1}^{n} \frac{\left(w^T x(i) - y(i)\right)^2}{n}$$

It is the sum, over all the data points, of the square of the difference between the predicted and actual target variables, divided by the number of data points.

RMSE is the square root of MSE. MSE is measured in units that are the square of the target variable, while RMSE is measured in the same units as the target variable. Due to its formulation, MSE, just like the squared loss function that it derives from, effectively penalizes larger errors more severely.

In order to evaluate our predictions based on the mean of an error metric, we will first make predictions for each input feature vector in an RDD of `LabeledPoint` instances by computing the error for each record using a function that takes the prediction and true target value as inputs. This will return a `[Double]` RDD that contains the error values. We can then find the average using the `mean` method of RDDs that contain `Double` values.

Let's define our squared error function as follows:

```
def squared_error(actual, pred):
    return (pred - actual)**2
```

Mean Absolute Error

MAE is the average of the absolute differences between the predicted and actual targets:

$$\sum_{i=1}^{n} \frac{|w^T x(i) - y(i)|}{n}$$

MAE is similar in principle to MSE, but it does not punish large deviations as much.

Our function to compute MAE is as follows:

```
def abs_error(actual, pred):
    return np.abs(pred - actual)
```

Root Mean Squared Log Error

This measurement is not as widely used as MSE and MAE, but it is used as the metric for the Kaggle competition that uses the bike sharing dataset. It is effectively the RMSE of the log-transformed predicted and target values. This measurement is useful when there is a wide range in the target variable, and you do not necessarily want to penalize large errors when the predicted and target values are themselves high. It is also effective when you care about percentage errors rather than the absolute value of errors.

> The Kaggle competition evaluation page can be found at https://www.kaggle.com/c/bike-sharing-demand/details/evaluation.

The function to compute RMSLE is shown here:

```
def squared_log_error(pred, actual):
    return (np.log(pred + 1) - np.log(actual + 1))**2
```

The R-squared coefficient

The R-squared coefficient, also known as the coefficient of determination, is a measure of how well a model fits a dataset. It is commonly used in statistics. It measures the degree of variation in the target variable; this is explained by the variation in the input features. An R-squared coefficient generally takes a value between 0 and 1, where 1 equates to a perfect fit of the model.

Computing performance metrics on the bike sharing dataset

Given the functions we defined earlier, we can now compute the various evaluation metrics on our bike sharing data.

Linear model

Our approach will be to apply the relevant error function to each record in the RDD we computed earlier, which is true_vs_predicted for our linear model:

```
mse = true_vs_predicted.map(lambda (t, p): squared_error(t, p)).mean()
mae = true_vs_predicted.map(lambda (t, p): abs_error(t, p)).mean()
rmsle = np.sqrt(true_vs_predicted.map(lambda (t, p): squared_log_error(t, p)).mean())
print "Linear Model - Mean Squared Error: %2.4f" % mse
print "Linear Model - Mean Absolute Error: %2.4f" % mae
print "Linear Model - Root Mean Squared Log Error: %2.4f" % rmsle
```

This outputs the following metrics:

```
Linear Model - Mean Squared Error: 28166.3824
Linear Model - Mean Absolute Error: 129.4506
Linear Model - Root Mean Squared Log Error: 1.4974
```

Decision tree

We will use the same approach for the decision tree model, using the `true_vs_predicted_dt` RDD:

```
mse_dt = true_vs_predicted_dt.map(lambda (t, p): squared_error(t, p)).mean()
mae_dt = true_vs_predicted_dt.map(lambda (t, p): abs_error(t, p)).mean()
rmsle_dt = np.sqrt(true_vs_predicted_dt.map(lambda (t, p): squared_log_error(t, p)).mean())
print "Decision Tree - Mean Squared Error: %2.4f" % mse_dt
print "Decision Tree - Mean Absolute Error: %2.4f" % mae_dt
print "Decision Tree - Root Mean Squared Log Error: %2.4f" % rmsle_dt
```

You should see output similar to this:

```
Decision Tree - Mean Squared Error: 11560.7978
Decision Tree - Mean Absolute Error: 71.0969
Decision Tree - Root Mean Squared Log Error: 0.6259
```

Looking at the results, we can see that our initial guess about the decision tree model being the better performer is indeed true.

> The Kaggle competition leaderboard lists the Mean Value Benchmark score on the test set at about 1.58. So, we see that our linear model performance is not much better. However, the decision tree with default settings achieves a performance of 0.63.
>
> The winning score at the time of writing this book is listed as 0.29504.

Improving model performance and tuning parameters

In *Chapter 5, Building a Classification Model with Spark*, we showed how feature transformation and selection can make a large difference to the performance of a model. In this chapter, we will focus on another type of transformation that can be applied to a dataset: transforming the target variable itself.

Transforming the target variable

Recall that many machine learning models, including linear models, make assumptions regarding the distribution of the input data as well as target variables. In particular, linear regression assumes a normal distribution.

In many real-world cases, the distributional assumptions of linear regression do not hold. In this case, for example, we know that the number of bike rentals can never be negative. This alone should indicate that the assumption of normality might be problematic. To get a better idea of the target distribution, it is often a good idea to plot a histogram of the target values.

In this section, if you are using IPython Notebook, enter the magic function, `%pylab inline`, to import `pylab` (that is, the `numpy` and `matplotlib` plotting functions) into the workspace. This will also create any figures and plots inline within the `Notebook` cell.

If you are using the standard IPython console, you can use `%pylab` to import the necessary functionality (your plots will appear in a separate window).

We will now create a plot of the target variable distribution in the following piece of code:

```
targets = records.map(lambda r: float(r[-1])).collect()
hist(targets, bins=40, color='lightblue', normed=True)
fig = matplotlib.pyplot.gcf()
fig.set_size_inches(16, 10)
```

Looking at the histogram plot, we can see that the distribution is highly skewed and certainly does not follow a normal distribution:

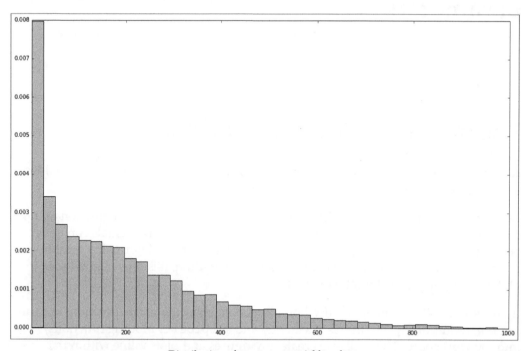

Distribution of raw target variable values

One way in which we might deal with this situation is by applying a transformation to the target variable, such that we take the logarithm of the target value instead of the raw value. This is often referred to as log-transforming the target variable (this transformation can also be applied to feature values).

We will apply a log transformation to the following target variable and plot a histogram of the log-transformed values:

```
log_targets = records.map(lambda r: np.log(float(r[-1]))).collect()
hist(log_targets, bins=40, color='lightblue', normed=True)
fig = matplotlib.pyplot.gcf()
fig.set_size_inches(16, 10)
```

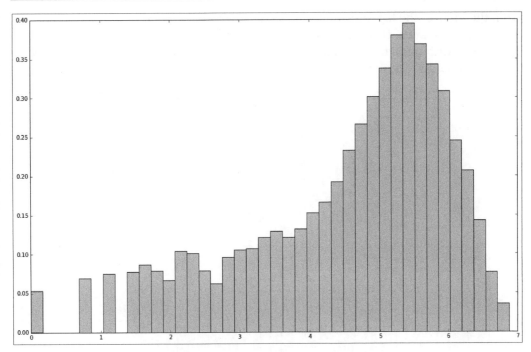

Distribution of log-transformed target variable values

A second type of transformation that is useful in the case of target values that do not take on negative values and, in addition, might take on a very wide range of values, is to take the square root of the variable.

We will apply the square root transform in the following code, once more plotting the resulting target variable distribution:

```
sqrt_targets = records.map(lambda r: np.sqrt(float(r[-1]))).collect()
hist(sqrt_targets, bins=40, color='lightblue', normed=True)
fig = matplotlib.pyplot.gcf()
fig.set_size_inches(16, 10)
```

From the plots of the log and square root transformations, we can see that both result in a more even distribution relative to the raw values. While they are still not normally distributed, they are a lot closer to a normal distribution when compared to the original target variable.

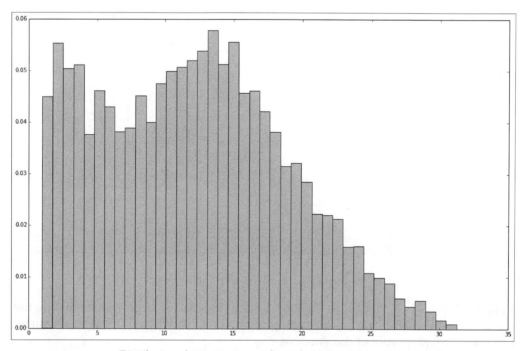

Distribution of square-root-transformed target variable values

Impact of training on log-transformed targets

So, does applying these transformations have any impact on model performance? Let's evaluate the various metrics we used previously on log-transformed data as an example.

We will do this first for the linear model by applying the numpy log function to the label field of each LabeledPoint RDD. Here, we will only transform the target variable, and we will not apply any transformations to the features:

```
data_log = data.map(lambda lp: LabeledPoint(np.log(lp.label),
lp.features))
```

We will then train a model on this transformed data and form the RDD of predicted versus true values:

```
model_log = LinearRegressionWithSGD.train(data_log, iterations=10, step=0.1)
```

Note that now that we have transformed the target variable, the predictions of the model will be on the log scale, as will the target values of the transformed dataset. Therefore, in order to use our model and evaluate its performance, we must first transform the log data back into the original scale by taking the exponent of both the predicted and true values using the `numpy exp` function. We will show you how to do this in the code here:

```
true_vs_predicted_log = data_log.map(lambda p: (np.exp(p.label), np.exp(model_log.predict(p.features))))
```

Finally, we will compute the MSE, MAE, and RMSLE metrics for the model:

```
mse_log = true_vs_predicted_log.map(lambda (t, p): squared_error(t, p)).mean()
mae_log = true_vs_predicted_log.map(lambda (t, p): abs_error(t, p)).mean()
rmsle_log = np.sqrt(true_vs_predicted_log.map(lambda (t, p): squared_log_error(t, p)).mean())
print "Mean Squared Error: %2.4f" % mse_log
print "Mean Absolue Error: %2.4f" % mae_log
print "Root Mean Squared Log Error: %2.4f" % rmsle_log
print "Non log-transformed predictions:\n" + str(true_vs_predicted.take(3))
print "Log-transformed predictions:\n" + str(true_vs_predicted_log.take(3))
```

You should see output similar to the following:

Mean Squared Error: 38606.0875

Mean Absolue Error: 135.2726

Root Mean Squared Log Error: 1.3516

Non log-transformed predictions:

`[(16.0, 119.30920003093594), (40.0, 116.95463511937378), (32.0, 116.57294610647752)]`

Log-transformed predictions:

`[(15.999999999999998, 45.860944832110015), (40.0, 43.255903592233274), (32.0, 42.311306147884252)]`

If we compare these results to the raw target variable, we see that while we did not improve the MSE or MAE, we improved the RMSLE.

We will perform the same analysis for the decision tree model:

```
data_dt_log = data_dt.map(lambda lp:
LabeledPoint(np.log(lp.label), lp.features))
dt_model_log = DecisionTree.trainRegressor(data_dt_log,{})

preds_log = dt_model_log.predict(data_dt_log.map(lambda p:
p.features))
actual_log = data_dt_log.map(lambda p: p.label)
true_vs_predicted_dt_log = actual_log.zip(preds_log).map(lambda (t,
p): (np.exp(t), np.exp(p)))

mse_log_dt = true_vs_predicted_dt_log.map(lambda (t, p): squared_
error(t, p)).mean()
mae_log_dt = true_vs_predicted_dt_log.map(lambda (t, p): abs_error(t,
p)).mean()
rmsle_log_dt = np.sqrt(true_vs_predicted_dt_log.map(lambda (t, p):
squared_log_error(t, p)).mean())
print "Mean Squared Error: %2.4f" % mse_log_dt
print "Mean Absolue Error: %2.4f" % mae_log_dt
print "Root Mean Squared Log Error: %2.4f" % rmsle_log_dt
print "Non log-transformed predictions:\n" + str(true_vs_predicted_
dt.take(3))
print "Log-transformed predictions:\n" +
str(true_vs_predicted_dt_log.take(3))
```

From the results here, we can see that we actually made our metrics slightly worse for the decision tree:

```
Mean Squared Error: 14781.5760
Mean Absolue Error: 76.4131
Root Mean Squared Log Error: 0.6406
Non log-transformed predictions:
[(16.0, 54.913223140495866), (40.0, 54.913223140495866), (32.0,
53.171052631578945)]
Log-transformed predictions:
[(15.999999999999998, 37.530779787154508), (40.0,
37.530779787154508), (32.0, 7.2797070993907287)]
```

It is probably not surprising that the log transformation results in a better RMSLE performance for the linear model. As we are minimizing the squared error, once we have transformed the target variable to log values, we are effectively minimizing a loss function that is very similar to the RMSLE.

This is good for Kaggle competition purposes, since we can more directly optimize against the competition-scoring metric.

It might or might not be as useful in a real-world situation. This depends on how important larger absolute errors are (recall that RMSLE essentially penalizes relative errors rather than absolute magnitude of errors).

Tuning model parameters

So far in this chapter, we have illustrated the concepts of model training and evaluation for MLlib's regression models by training and testing on the same dataset. We will now use a similar cross-validation approach that we used previously to evaluate the effect on performance of different parameter settings for our models.

Creating training and testing sets to evaluate parameters

The first step is to create a test and training set for cross-validation purposes. Spark's Python API does not yet provide the `randomSplit` convenience method that is available in Scala. Hence, we will need to create a training and test dataset manually.

One relatively easy way to do this is by first taking a random sample of, say, 20 percent of our data as our test set. We will then define our training set as the elements of the original RDD that are not in the test set RDD.

We can achieve this using the `sample` method to take a random sample for our test set, followed by using the `subtractByKey` method, which takes care of returning the elements in one RDD where the keys do not overlap with the other RDD.

Note that `subtractByKey`, as the name suggests, works on the keys of the RDD elements that consist of key-value pairs. Therefore, here we will use `zipWithIndex` on our RDD of extracted training examples. This creates an RDD of `(LabeledPoint, index)` pairs.

We will then reverse the keys and values so that we can operate on the index keys:

```
data_with_idx = data.zipWithIndex().map(lambda (k, v): (v, k))
test = data_with_idx.sample(False, 0.2, 42)
train = data_with_idx.subtractByKey(test)
```

Once we have the two RDDs, we will recover just the `LabeledPoint` instances we need for training and test data, using map to extract the value from the key-value pairs:

```
train_data = train.map(lambda (idx, p): p)
test_data = test.map(lambda (idx, p) : p)
train_size = train_data.count()
test_size = test_data.count()
print "Training data size: %d" % train_size
print "Test data size: %d" % test_size
print "Total data size: %d " % num_data
print "Train + Test size : %d" % (train_size + test_size)
```

We can confirm that we now have two distinct datasets that add up to the original dataset in total:

```
Training data size: 13934
Test data size: 3445
Total data size: 17379
Train + Test size : 17379
```

The final step is to apply the same approach to the features extracted for the decision tree model:

```
data_with_idx_dt = data_dt.zipWithIndex().map(lambda (k, v): (v, k))
test_dt = data_with_idx_dt.sample(False, 0.2, 42)
train_dt = data_with_idx_dt.subtractByKey(test_dt)
train_data_dt = train_dt.map(lambda (idx, p): p)
test_data_dt = test_dt.map(lambda (idx, p) : p)
```

The impact of parameter settings for linear models

Now that we have prepared our training and test sets, we are ready to investigate the impact of different parameter settings on model performance. We will first carry out this evaluation for the linear model. We will create a convenience function to evaluate the relevant performance metric by training the model on the training set and evaluating it on the test set for different parameter settings.

We will use the RMSLE evaluation metric, as it is the one used in the Kaggle competition with this dataset, and this allows us to compare our model results against the competition leaderboard to see how we perform.

The evaluation function is defined here:

```
def evaluate(train, test, iterations, step, regParam, regType,
intercept):
    model = LinearRegressionWithSGD.train(train, iterations, step,
regParam=regParam, regType=regType, intercept=intercept)
    tp = test.map(lambda p: (p.label, model.predict(p.features)))
    rmsle = np.sqrt(tp.map(lambda (t, p): squared_log_error(t, p)).mean())
    return rmsle
```

 Note that in the following sections, you might get slightly different results due to some random initialization for SGD. However, your results will be comparable.

Iterations

As we saw when evaluating our classification models, we generally expect that a model trained with SGD will achieve better performance as the number of iterations increases, although the increase in performance will slow down as the number of iterations goes above some minimum number. Note that here, we will set the step size to 0.01 to better illustrate the impact at higher iteration numbers:

```
params = [1, 5, 10, 20, 50, 100]
metrics = [evaluate(train_data, test_data, param, 0.01, 0.0, 'l2',
False) for param in params]
print params
print metrics
```

The output shows that the error metric indeed decreases as the number of iterations increases. It also does so at a decreasing rate, again as expected. What is interesting is that eventually, the SGD optimization tends to overshoot the optimal solution, and the RMSLE eventually starts to increase slightly:

```
[1, 5, 10, 20, 50, 100]
[2.3532904530306888, 1.6438528499254723, 1.4869656275309227,
1.4149741941240344, 1.4159641262731959, 1.4539667094611679]
```

Building a Regression Model with Spark

Here, we will use the matplotlib library to plot a graph of the RMSLE metric against the number of iterations. We will use a log scale for the *x* axis to make the output easier to visualize:

```
plot(params, metrics)
fig = matplotlib.pyplot.gcf()
pyplot.xscale('log')
```

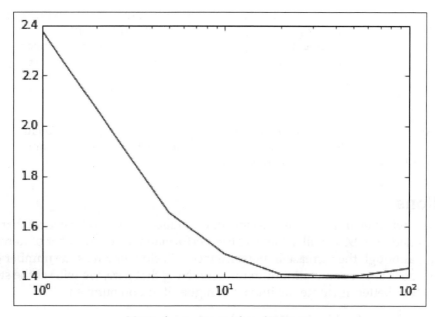

Metrics for varying number of iterations

Step size

We will perform a similar analysis for step size in the following code:

```
params = [0.01, 0.025, 0.05, 0.1, 1.0]
metrics = [evaluate(train_data, test_data, 10, param, 0.0, 'l2',
False) for param in params]
print params
print metrics
```

The output of the preceding code:

```
[0.01, 0.025, 0.05, 0.1, 0.5]

[1.4869656275309227, 1.4189071944747715, 1.5027293911925559,
1.5384660954019973, nan]
```

Now, we can see why we avoided using the default step size when training the linear model originally. The default is set to *1.0*, which, in this case, results in a nan output for the RMSLE metric. This typically means that the SGD model has converged to a very poor local minimum in the error function that it is optimizing. This can happen when the step size is relatively large, as it is easier for the optimization algorithm to overshoot good solutions.

We can also see that for low step sizes and a relatively low number of iterations (we used 10 here), the model performance is slightly poorer. However, in the preceding *Iterations* section, we saw that for the lower step-size setting, a higher number of iterations will generally converge to a better solution.

Generally speaking, setting step size and number of iterations involves a trade-off. A lower step size means that convergence is slower but slightly more assured. However, it requires a higher number of iterations, which is more costly in terms of computation and time, in particular at a very large scale.

> Selecting the best parameter settings can be an intensive process that involves training a model on many combinations of parameter settings and selecting the best outcome. Each instance of model training involves a number of iterations, so this process can be very expensive and time consuming when performed on very large datasets.

The output is plotted here, again using a log scale for the step-size axis:

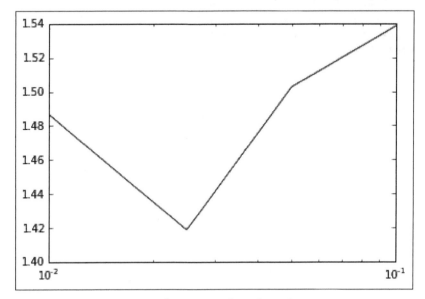

Metrics for varying values of step size

L2 regularization

In *Chapter 5, Building a Classification Model with Spark*, we saw that regularization has the effect of penalizing model complexity in the form of an additional loss term that is a function of the model weight vector. L2 regularization penalizes the L2-norm of the weight vector, while L1 regularization penalizes the L1-norm.

We expect training set performance to deteriorate with increasing regularization, as the model cannot fit the dataset well. However, we would also expect some amount of regularization that will result in optimal generalization performance as evidenced by the best performance on the test set.

We will evaluate the impact of different levels of L2 regularization in this code:

```
params = [0.0, 0.01, 0.1, 1.0, 5.0, 10.0, 20.0]
metrics = [evaluate(train_data, test_data, 10, 0.1, param, 'l2',
False) for param in params]
print params
print metrics
plot(params, metrics)
fig = matplotlib.pyplot.gcf()
pyplot.xscale('log')
```

As expected, there is an optimal setting of the regularization parameter with respect to the test set RMSLE:

```
[0.0, 0.01, 0.1, 1.0, 5.0, 10.0, 20.0]
[1.5384660954019971, 1.5379108106882864, 1.5329809395123755,
1.4900275345312988, 1.4016676336981468, 1.40998359211149,
1.5381771283158705]
```

This is easiest to see in the following plot (where we once more use the log scale for the regularization parameter axis):

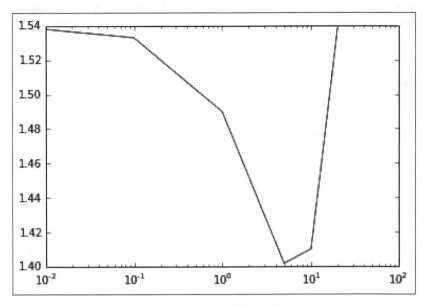

Metrics for varying levels of L2 regularization

L1 regularization

We can apply the same approach for differing levels of L1 regularization:

```
params = [0.0, 0.01, 0.1, 1.0, 10.0, 100.0, 1000.0]
metrics = [evaluate(train_data, test_data, 10, 0.1, param, 'l1',
False) for param in params]
print params
print metrics
plot(params, metrics)
fig = matplotlib.pyplot.gcf()
pyplot.xscale('log')
```

Again, the results are more clearly seen when plotted in the following graph. We see that there is a much more subtle decline in RMSLE, and it takes a very high value to cause a jump back up. Here, the level of L1 regularization required is much higher than that for the L2 form; however, the overall performance is poorer:

```
[0.0, 0.01, 0.1, 1.0, 10.0, 100.0, 1000.0]
[1.5384660954019971, 1.5384518080419873, 1.5383237472930684,
1.5372017600929164, 1.5303809928601677, 1.4352494587433793,
4.7551250073268614]
```

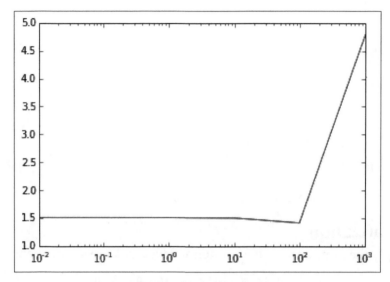

Metrics for varying levels of L1 regularization

Using L1 regularization can encourage sparse weight vectors. Does this hold true in this case? We can find out by examining the number of entries in the weight vector that are zero, with increasing levels of regularization:

```
model_l1 = LinearRegressionWithSGD.train(train_data, 10, 0.1,
regParam=1.0, regType='l1', intercept=False)
model_l1_10 = LinearRegressionWithSGD.train(train_data, 10, 0.1,
regParam=10.0, regType='l1', intercept=False)
model_l1_100 = LinearRegressionWithSGD.train(train_data, 10, 0.1,
regParam=100.0, regType='l1', intercept=False)
print "L1 (1.0) number of zero weights: " + str(sum(model_l1.weights.
array == 0))
```

```
print "L1 (10.0) number of zeros weights: " + str(sum(model_l1_10.
weights.array == 0))
print "L1 (100.0) number of zeros weights: " +
str(sum(model_l1_100.weights.array == 0))
```

We can see from the results that as we might expect, the number of zero feature weights in the model weight vector increases as greater levels of L1 regularization are applied:

L1 (1.0) number of zero weights: 4

L1 (10.0) number of zeros weights: 20

L1 (100.0) number of zeros weights: 55

Intercept

The final parameter option for the linear model is whether to use an intercept or not. An intercept is a constant term that is added to the weight vector and effectively accounts for the mean value of the target variable. If the data is already centered or normalized, an intercept is not necessary; however, it often does not hurt to use one in any case.

We will evaluate the effect of adding an intercept term to the model here:

```
params = [False, True]
metrics = [evaluate(train_data, test_data, 10, 0.1, 1.0, 'l2', param)
for param in params]
print params
print metrics
bar(params, metrics, color='lightblue')
fig = matplotlib.pyplot.gcf()
```

We can see from the result and plot that adding the intercept term results in a very slight increase in RMSLE:

```
[False, True]
```
```
[1.4900275345312988, 1.506469812020645]
```

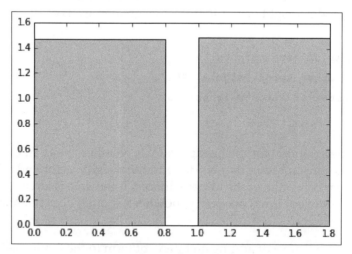

Metrics without and with an intercept

The impact of parameter settings for the decision tree

Decision trees provide two main parameters: maximum tree depth and the maximum number of bins. We will now perform the same evaluation of the effect of parameter settings for the decision tree model. Our starting point is to create an evaluation function for the model, similar to the one used for the linear regression earlier. This function is provided here:

```
def evaluate_dt(train, test, maxDepth, maxBins):
    model = DecisionTree.trainRegressor(train, {},
    impurity='variance', maxDepth=maxDepth, maxBins=maxBins)
    preds = model.predict(test.map(lambda p: p.features))
    actual = test.map(lambda p: p.label)
    tp = actual.zip(preds)
    rmsle = np.sqrt(tp.map(lambda (t, p): squared_log_error(t,
    p)).mean())
    return rmsle
```

Tree depth

We would generally expect performance to increase with more complex trees (that is, trees of greater depth). Having a lower tree depth acts as a form of regularization, and it might be the case that as with L2 or L1 regularization in linear models, there is a tree depth that is optimal with respect to the test set performance.

Here, we will try to increase the depths of trees to see what impact they have on test set RMSLE, keeping the number of bins at the default level of 32:

```
params = [1, 2, 3, 4, 5, 10, 20]
metrics = [evaluate_dt(train_data_dt, test_data_dt, param, 32) for param in params]
print params
print metrics
plot(params, metrics)
fig = matplotlib.pyplot.gcf()
```

In this case, it appears that the decision tree starts over-fitting at deeper tree levels. An optimal tree depth appears to be around 10 on this dataset.

 Notice that our best RMSLE of 0.42 is now quite close to the Kaggle winner of around 0.29!

The output of the tree depth is as follows:

[1, 2, 3, 4, 5, 10, 20]

[1.0280339660196287, 0.92686672078778276, 0.81807794023407532, 0.74060228537329209, 0.63583503599563096, 0.42851360418692447, 0.45500008049779139]

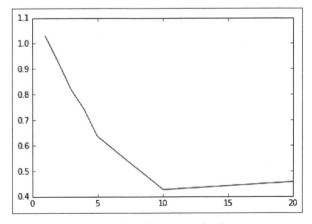

Metrics for different tree depths

Maximum bins

Finally, we will perform our evaluation on the impact of setting the number of bins for the decision tree. As with the tree depth, a larger number of bins should allow the model to become more complex and might help performance with larger feature dimensions. After a certain point, it is unlikely that it will help any more and might, in fact, hinder performance on the test set due to over-fitting:

```
params = [2, 4, 8, 16, 32, 64, 100]
metrics = [evaluate_dt(train_data_dt, test_data_dt, 5, param) for
param in params]
print params
print metrics
plot(params, metrics)
fig = matplotlib.pyplot.gcf()
```

Here, we will show the output and plot to vary the number of bins (while keeping the tree depth at the default level of 5). In this case, using a small number of bins hurts performance, while there is no impact when we use around 32 bins (the default setting) or more. There seems to be an optimal setting for test set performance at around 16-20 bins:

[2, 4, 8, 16, 32, 64, 100]

[1.3069788763726049, 0.81923394899750324, 0.75745322513058744, 0.62328384445223795, 0.63583503599563096, 0.63583503599563096, 0.63583503599563096]

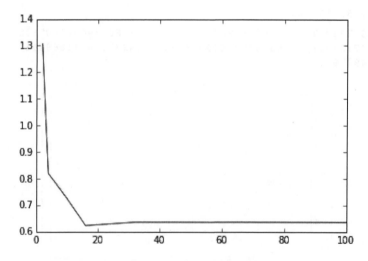

Metrics for different maximum bins

Summary

In this chapter, you saw how to use MLlib's linear model and decision tree functionality in Python within the context of regression models. We explored categorical feature extraction and the impact of applying transformations to the target variable in a regression problem. Finally, we implemented various performance-evaluation metrics and used them to implement a cross-validation exercise that explores the impact of the various parameter settings available in both linear models and decision trees on test set model performance.

In the next chapter, we will cover a different approach to machine learning, that is unsupervised learning, specifically in clustering models.

Building a Clustering Model with Spark

In the last few chapters, we covered supervised learning methods, where the training data is labeled with the true outcome that we would like to predict (for example, a rating for recommendations and class assignment for classification or real target variable in the case of regression).

Next, we will consider the case when we do not have labeled data available. This is called unsupervised learning, as the model is not supervised with the true target label. The unsupervised case is very common in practice, since obtaining labeled training data can be very difficult or expensive in many real-world scenarios (for example, having humans label training data with class labels for classification). However, we would still like to learn some underlying structure in the data and use these to make predictions.

This is where unsupervised learning approaches can be useful. Unsupervised learning models are also often combined with supervised models, for example, applying unsupervised techniques to create new input features for supervised models.

Clustering models are, in many ways, the unsupervised equivalent of classification models. With classification, we tried to learn a model that would predict which class a given training example belonged to. The model was essentially a mapping from a set of features to the class.

In clustering, we would like to segment the data such that each training example is assigned to a segment called a **cluster**. The clusters act much like classes, except that the true class assignments are unknown.

Clustering models have many use cases that are the same as classification; these include the following:

- Segmenting users or customers into different groups based on behavior characteristics and metadata
- Grouping content on a website or products in a retail business
- Finding clusters of similar genes
- Segmenting communities in ecology
- Creating image segments for use in image analysis applications such as object detection

In this chapter, we will:

- Briefly explore a few types of clustering models
- Extract features from data specifically using the output of one model as input features for our clustering model
- Train a clustering model and use it to make predictions
- Apply performance-evaluation and parameter-selection techniques to select the optimal number of clusters to use

Types of clustering models

There are many different forms of clustering models available, ranging from simple to extremely complex ones. The MLlib library currently provides K-means clustering, which is among the simplest approaches available. However, it is often very effective, and its simplicity means it is relatively easy to understand and is scalable.

K-means clustering

K-means attempts to partition a set of data points into K distinct clusters (where K is an input parameter for the model).

More formally, K-means tries to find clusters so as to minimize the sum of squared errors (or distances) within each cluster. This objective function is known as the **within cluster sum of squared errors (WCSS)**.

$$\sum_{i=1}^{n} \sum_{j=1}^{n} (x(j) - u(i))^2$$

It is the sum, over each cluster, of the squared errors between each point and the cluster center.

Starting with a set of K initial cluster centers (which are computed as the mean vector for all data points in the cluster), the standard method for K-means iterates between two steps:

1. Assign each data point to the cluster that minimizes the WCSS. The sum of squares is equivalent to the squared Euclidean distance; therefore, this equates to assigning each point to the **closest** cluster center as measured by the Euclidean distance metric.
2. Compute the new cluster centers based on the cluster assignments from the first step.

The algorithm proceeds until either a maximum number of iterations has been reached or **convergence** has been achieved. Convergence means that the cluster assignments no longer change during the first step; therefore, the value of the WCSS objective function does not change either.

> For more details, refer to Spark's documentation on clustering at http://spark.apache.org/docs/latest/mllib-clustering.html or refer to http://en.wikipedia.org/wiki/K-means_clustering.

To illustrate the basics of K-means, we will use the simple dataset we showed in our multiclass classification example in *Chapter 5, Building a Classification Model with Spark*. Recall that we have five classes, which are shown in the following figure:

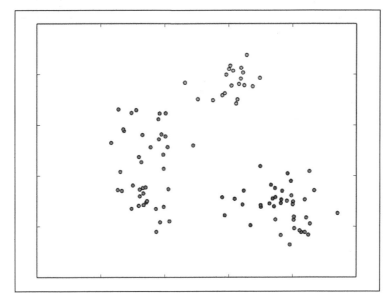

Multiclass dataset

However, assume that we don't actually know the true classes. If we use K-means with five clusters, then after the first step, the model's cluster assignments might look like this:

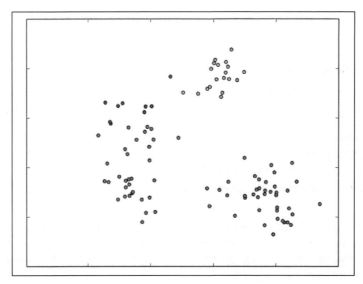

Cluster assignments after the first K-means iteration

We can see that K-means has already picked out the centers of each cluster fairly well. After the next iteration, the assignments might look like those shown in the following figure:

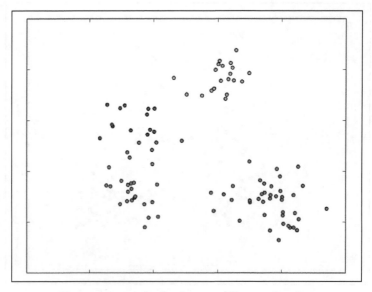

Cluster assignments after the second K-means iteration

Things are starting to stabilize, but the overall cluster assignments are broadly the same as they were after the first iteration. Once the model has converged, the final assignments could look like this:

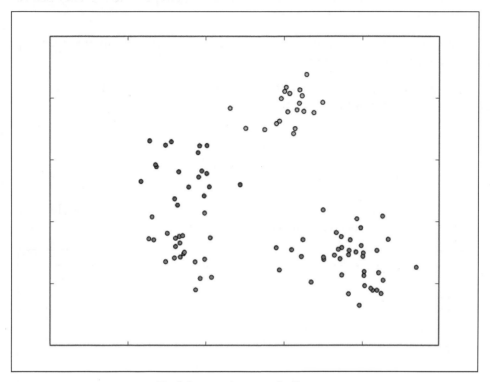

Final cluster assignments for K-means

As we can see, the model has done a decent job of separating the five clusters. The leftmost three are fairly accurate (with a few incorrect points). However, the two clusters in the bottom-right corner are less accurate.

This illustrates:

- The iterative nature of K-means
- The model's dependency on the method of initially selecting clusters' centers (here, we will use a random approach)
- That the final cluster assignments can be very good for well-separated data but can be poor for data that is more difficult

Initialization methods

The standard initialization method for K-means, usually simply referred to as the random method, starts by randomly assigning each data point to a cluster before proceeding with the first update step.

MLlib provides a parallel variant for this initialization method, called K-means ||, which is the default initialization method used.

MLlib provides a parallel variant called **K-means ||, ||**, for this initialization method; this is the default initialization method used.

> See http://en.wikipedia.org/wiki/K-means_clustering#Initialization_methods and http://en.wikipedia.org/wiki/K-means%2B%2B for more information.

The results of using K-means++ are shown here. Note that this time, the difficult lower-right points have been mostly correctly clustered.

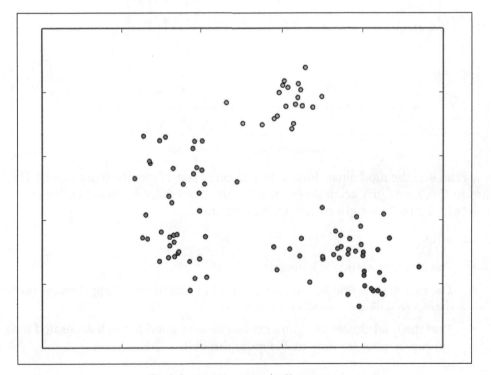

Final cluster assignments for K-means++

Variants

There are many other variants of K-means; they focus on initialization methods or the core model. One of the more common variants is fuzzy K-means. This model does not assign each point to one cluster as K-means does (a so-called hard assignment). Instead, it is a soft version of K-means, where each point can belong to many clusters, and is represented by the relative membership to each cluster. So, for K clusters, each point is represented as a K-dimensional membership vector, with each entry in this vector indicating the membership proportion in each cluster.

Mixture models

A **mixture model** is essentially an extension of the idea behind fuzzy K-means; however, it makes an assumption that there is an underlying probability distribution that generates the data. For example, we might assume that the data points are drawn from a set of K-independent Gaussian (normal) probability distributions. The cluster assignments are also soft, so each point is represented by K membership weights in each of the K underlying probability distributions.

See http://en.wikipedia.org/wiki/Mixture_model for further details and for a mathematical treatment of mixture models.

Hierarchical clustering

Hierarchical clustering is a structured clustering approach that results in a multilevel hierarchy of clusters, where each cluster might contain many subclusters (or child clusters). Each child cluster is, thus, linked to the parent cluster. This form of clustering is often also called tree clustering.

Agglomerative clustering is a bottom-up approach where:

- Each data point begins in its own cluster
- The similarity (or distance) between each pair of clusters is evaluated
- The pair of clusters that are most similar are found; this pair is then merged to form a new cluster
- The process is repeated until only one top-level cluster remains

Divisive clustering is a top-down approach that works in reverse, starting with one cluster and at each stage, splitting a cluster into two, until all data points are allocated to their own bottom-level cluster.

You can find more information at http://en.wikipedia.org/wiki/Hierarchical_clustering.

Extracting the right features from your data

Like most of the machine learning models we have encountered so far, K-means clustering requires numerical vectors as input. The same feature extraction and transformation approaches that we have seen for classification and regression are applicable for clustering.

As K-means, like least squares regression, uses a squared error function as the optimization objective, it tends to be impacted by outliers and features with large variance.

As for regression and classification cases, input data can be normalized and standardized to overcome this, which might improve accuracy. In some cases, however, it might be desirable not to standardize data, if, for example, the objective is to find segmentations according to certain specific features.

Extracting features from the MovieLens dataset

For this example, we will return to the movie rating dataset we used in *Chapter 4*, *Building a Recommendation Engine with Spark*. Recall that we have three main datasets: one that contains the movie ratings (in the u.data file), a second one with user data (u.user), and a third one with movie data (u.item). We will also be using the genre data file to extract the genres for each movie (u.genre).

We will start by looking at the movie data:

```
val movies = sc.textFile("/PATH/ml-100k/u.item")
println(movies.first)
```

This should output the first line of the dataset:

`1|Toy Story (1995)|01-Jan-1995||http://us.imdb.com/M/title-exact?Toy%20Story%20(1995)|0|0|0|1|1|1|0|0|0|0|0|0|0|0|0|0|0|0|0`

So, we have access to the move title, and we already have the movies categorized into genres. Why do we need to apply a clustering model to the movies? Clustering the movies is a useful exercise for two reasons:

- First, because we have access to the true genre labels, we can use these to evaluate the quality of the clusters that the model finds
- Second, we might wish to segment the movies based on some other attributes or features, apart from their genres

For example, in this case, it seems that we don't have a lot of data to use for clustering, apart from the genres and title. However, this is not true—we also have the ratings data. Previously, we created a matrix factorization model from the ratings data. The model is made up of a set of user and movie factor vectors.

We can think of the movie factors as representing each movie in a new latent feature space, where each latent feature, in turn, represents some form of structure in the ratings matrix. While it is not possible to directly interpret each latent feature, they might represent some hidden structure that influences the ratings behavior between users and movies. One factor could represent genre preference, another could refer to actors or directors, while yet another could represent the theme of the movie, and so on.

So, if we use these factor vector representations of each movie as inputs to our clustering model, we will end up with a clustering that is based on the actual rating behavior of users rather than manual genre assignments.

The same logic applies to the user factors—they represent users in the latent feature space of rating behavior, so clustering the user vectors should result in a clustering based on user rating behavior.

Extracting movie genre labels

Before proceeding further, let's extract the genre mappings from the u.genre file. As you can see from the first line of the preceding dataset, we will need to map from the numerical genre assignments to the textual version so that they are readable.

Take a look at the first few lines of u.genre:

```
val genres = sc.textFile("/PATH/ml-100k/u.genre")
genres.take(5).foreach(println)
```

You should see the following output displayed:

unknown|0

Action|1

Adventure|2

Animation|3

Children's|4

Here, 0 is the index of the relevant genre, while unknown is the genre assigned for this index. The indices correspond to the indices of the binary subvector that will represent the genres for each movie (that is, the 0s and 1s in the preceding movie data).

To extract the genre mappings, we will split each line and extract a key-value pair, where the key is the text genre and the value is the index. Note that we have to filter out an empty line at the end; this will, otherwise, throw an error when we try to split the line (see the code highlighted here):

```
val genreMap = genres.filter(!_.isEmpty).map(line => line.
split("\\|")).map(array => (array(1), array(0))).collectAsMap
println(genreMap)
```

The preceding code will provide the following output:

```
Map(2 -> Adventure, 5 -> Comedy, 12 -> Musical, 15 -> Sci-Fi, 8 -> Drama,
18 -> Western, ...
```

Next, we'll create a new RDD from the movie data and our genre mapping; this RDD contains the movie ID, title, and genres. We will use this later to create a more readable output when we evaluate the clusters assigned to each movie by our clustering model.

In the following code section, we will map over each movie and extract the genres subvector (which will still contain Strings rather than Int indexes). We will then apply the zipWithIndex method to create a new collection that contains the indices of the genre subvector, and we will filter this collection so that we are left only with the positive assignments (that is, the 1s that denote a genre assignment for the relevant index). We can then use our extracted genre mapping to map these indices to the textual genres. Finally, we will inspect the first record of the new RDD to see the result of these operations:

```
val titlesAndGenres = movies.map(_.split("\\|")).map { array =>
  val genres = array.toSeq.slice(5, array.size)
  val genresAssigned = genres.zipWithIndex.filter { case (g, idx)
=>
    g == "1"
  }.map { case (g, idx) =>
    genreMap(idx.toString)
  }
  (array(0).toInt, (array(1), genresAssigned))
}
println(titlesAndGenres.first)
```

This should output the following result:

```
(1,(Toy Story (1995),ArrayBuffer(Animation, Children's, Comedy)))
```

Training the recommendation model

To get the user and movie factor vectors, we first need to train another recommendation model. Fortunately, we have already done this in *Chapter 4, Building a Recommendation Engine with Spark*, so we will follow the same procedure:

```
import org.apache.spark.mllib.recommendation.ALS
import org.apache.spark.mllib.recommendation.Rating
val rawData = sc.textFile("/PATH/ml-100k/u.data")
val rawRatings = rawData.map(_.split("\t").take(3))
val ratings = rawRatings.map{ case Array(user, movie, rating) =>
Rating(user.toInt, movie.toInt, rating.toDouble) }
ratings.cache
val alsModel = ALS.train(ratings, 50, 10, 0.1)
```

Recall from *Chapter 4, Building a Recommendation Engine with Spark*, that the ALS model returned contains the factors in two RDDs of key-value pairs (called `userFeatures` and `productFeatures`) with the user or movie ID as the key and the factor as the value. We will need to extract just the factors and transform each one of them into an MLlib `Vector` to use as training input for our clustering model.

We will do this for both users and movies as follows:

```
import org.apache.spark.mllib.linalg.Vectors
val movieFactors = alsModel.productFeatures.map { case (id, factor) =>
(id, Vectors.dense(factor)) }
val movieVectors = movieFactors.map(_._2)
val userFactors = alsModel.userFeatures.map { case (id, factor) =>
(id, Vectors.dense(factor)) }
val userVectors = userFactors.map(_._2)
```

Normalization

Before we train our clustering model, it might be useful to look into the distribution of the input data in the form of the factor vectors. This will tell us whether we need to normalize the training data.

We will follow the same approach as we did in *Chapter 5, Building a Classification Model with Spark*, using MLlib's summary statistics available in the distributed `RowMatrix` class:

```
import org.apache.spark.mllib.linalg.distributed.RowMatrix
val movieMatrix = new RowMatrix(movieVectors)
val movieMatrixSummary =
movieMatrix.computeColumnSummaryStatistics()
val userMatrix = new RowMatrix(userVectors)
val userMatrixSummary =
```

```
userMatrix.computeColumnSummaryStatistics()
println("Movie factors mean: " + movieMatrixSummary.mean)
println("Movie factors variance: " + movieMatrixSummary.variance)
println("User factors mean: " + userMatrixSummary.mean)
println("User factors variance: " + userMatrixSummary.variance)
```

You should see output similar to the one here:

```
Movie factors mean: [0.28047737659519767,0.26886479057520024,0.2935579964
446398,0.27821738264113755, ...
Movie factors variance: [0.038242041794064895,0.03742229118854288,0.04411
6961097355877,0.057116244055791986, ...
User factors mean: [0.2043520841572601,0.22135773814655782,0.214970631841
8221,0.23647602029329481, ...
User factors variance: [0.037749421148850396,0.02831191551960241,0.032831
876953314174,0.036775110657850954, ...
```

If we look at the output, we will see that there do not appear to be any important outliers that might skew the clustering results, so normalization should not be required in this case.

Training a clustering model

Training for K-means in MLlib takes an approach similar to the other models — we pass an RDD that contains our training data to the `train` method of the `KMeans` object. Note that here we do not use `LabeledPoint` instances, as the labels are not used in clustering; they are used only in the feature vectors. Thus, we use a RDD `[Vector]` as input to the `train` method.

Training a clustering model on the MovieLens dataset

We will train a model for both the movie and user factors that we generated by running our recommendation model. We need to pass in the number of clusters K and the maximum number of iterations for the algorithm to run. Model training might run for less than the maximum number of iterations if the change in the objective function from one iteration to the next is less than the tolerance level (the default for this tolerance is 0.0001).

MLlib's K-means provides random and K-means || initialization, with the default being K-means ||. As both of these initialization methods are based on random selection to some extent, each model training run will return a different result.

K-means does not generally converge to a global optimum model, so performing multiple training runs and selecting the most optimal model from these runs is a common practice. MLlib's training methods expose an option to complete multiple model training runs. The best training run, as measured by the evaluation of the loss function, is selected as the final model.

We will first set up the required imports, as well as model parameters: K, maximum iterations, and number of runs:

```
import org.apache.spark.mllib.clustering.KMeans
val numClusters = 5
val numIterations = 10
val numRuns = 3
```

We will then run K-means on the movie factor vectors:

```
val movieClusterModel = KMeans.train(movieVectors, numClusters, numIterations, numRuns)
```

Once the model has completed training, we should see output that looks something like this:

```
...
14/09/02 21:53:58 INFO SparkContext: Job finished: collectAsMap at KMeans.scala:193, took 0.02043 s
14/09/02 21:53:58 INFO KMeans: Iterations took 0.331 seconds.
14/09/02 21:53:58 INFO KMeans: KMeans reached the max number of iterations: 10.
14/09/02 21:53:58 INFO KMeans: The cost for the best run is 2586.298785925147.
...
movieClusterModel: org.apache.spark.mllib.clustering.KMeansModel = org.apache.spark.mllib.clustering.KMeansModel@71c6f512
```

As can be seen from the highlighted text, the model training output tells us that the maximum number of iterations was reached, so the training process did not stop early based on the convergence criterion. It also shows the training set error (that is, the value of the K-means objective function) for the best run.

We can try a much larger setting for the maximum iterations and use only one training run to see an example where the K-means model converges:

```
val movieClusterModelConverged = KMeans.train(movieVectors, numClusters, 100)
```

You should be able to see the `KMeans converged in ... iterations` text in the model output; this text indicates that after so many iterations, the K-means objective function did not decrease more than the tolerance level:

```
...
14/09/02 22:04:38 INFO SparkContext: Job finished: collectAsMap at
KMeans.scala:193, took 0.040685 s
14/09/02 22:04:38 INFO KMeans: Run 0 finished in 34 iterations
14/09/02 22:04:38 INFO KMeans: Iterations took 0.812 seconds.
14/09/02 22:04:38 INFO KMeans: KMeans converged in 34 iterations.
14/09/02 22:04:38 INFO KMeans: The cost for the best run is
2584.9354332904104.
...
movieClusterModelConverged: org.apache.spark.mllib.clustering.KMeansModel
= org.apache.spark.mllib.clustering.KMeansModel@6bb28fb5
```

> Notice that when we use a lower number of iterations but use multiple training runs, we typically get a training error (called cost above) that is very similar to the one we obtain by running the model to convergence. Using the multiple runs option can, therefore, be a very effective method to find the best possible model.

Finally, we will also train a K-means model on the user factor vectors:

```
val userClusterModel = KMeans.train(userVectors, numClusters,
numIterations, numRuns)
```

Making predictions using a clustering model

Using the trained K-means model is straightforward and similar to the other models we have encountered so far, such as classification and regression. We can make a prediction for a single `Vector` instance as follows:

```
val movie1 = movieVectors.first
val movieCluster = movieClusterModel.predict(movie1)
println(movieCluster)
```

We can also make predictions for multiple inputs by passing a RDD [Vector] to the predict method of the model:

```
val predictions = movieClusterModel.predict(movieVectors)
println(predictions.take(10).mkString(","))
```

The resulting output is a cluster assignment for each data point:

`0,0,1,1,2,1,0,1,1,1`

> Note that due to random initialization, the cluster assignments might change from one run of the model to another, so your results might differ from those shown earlier. The cluster ID themselves have no inherent meaning; they are simply arbitrarily labeled, starting from 0.

Interpreting cluster predictions on the MovieLens dataset

We have covered how to make predictions for a set of input vectors, but how do we evaluate how good the predictions are? We will cover performance metrics a little later; however, here, we will see how to manually inspect and interpret the cluster assignments made by our K-means model.

While unsupervised techniques have the advantage that they do not require us to provide labeled data for training, the disadvantage is that often, the results need to be manually interpreted. Often, we would like to further examine the clusters that are found and possibly try to interpret them and assign some sort of labeling or categorization to them.

For example, we can examine the clustering of movies we have found to try to see whether there is some meaningful interpretation of each cluster, such as a common genre or theme among the movies in the cluster. There are many approaches we can use, but we will start by taking a few movies in each cluster that are closest to the center of the cluster. These movies, we assume, would be the ones that are least likely to be marginal in terms of their cluster assignment, and so, they should be among the most representative of the movies in the cluster. By examining these sets of movies, we can see what attributes are shared by the movies in each cluster.

Interpreting the movie clusters

To begin, we need to decide what we mean by "closest to the center of each cluster". The objective function that is minimized by K-means is the sum of Euclidean distances between each point and the cluster center, summed over all clusters. Therefore, it is natural to use the Euclidean distance as our measure.

We will define this function here. Note that we will need access to certain imports from the **Breeze** library (a dependency of MLlib) for linear algebra and vector-based numerical functions:

```
import breeze.linalg._
import breeze.numerics.pow
def computeDistance(v1: DenseVector[Double], v2: DenseVector[Double])
= pow(v1 - v2, 2).sum
```

> The preceding pow function is a Breeze universal function. This function is the same as the pow function from scala.math, except that it operates element-wise on the vector that is returned from the minus operation between the two input vectors.

Now, we will use this function to compute, for each movie, the distance of the relevant movie factor vector from the center vector of the assigned cluster. We will also join our cluster assignments and distances data with the movie titles and genres so that we can output the results in a more readable way:

```
val titlesWithFactors = titlesAndGenres.join(movieFactors)
val moviesAssigned = titlesWithFactors.map { case (id, ((title,
genres), vector)) =>
  val pred = movieClusterModel.predict(vector)
  val clusterCentre = movieClusterModel.clusterCenters(pred)
  val dist = computeDistance(DenseVector(clusterCentre.toArray),
DenseVector(vector.toArray))
  (id, title, genres.mkString(" "), pred, dist)
}
val clusterAssignments = moviesAssigned.groupBy { case (id, title,
genres, cluster, dist) => cluster }.collectAsMap
```

After running the preceding code snippet, we have an RDD that contains a set of key-value pairs for each cluster; here, the key is the numeric cluster identifier, and the value is made up of a set of movies and related information. The movie information we have is the movie ID, title, genres, cluster index, and distance of the movie's factor vector from the cluster center.

Finally, we will iterate through each cluster and output the top 20 movies, ranked by distance from closest to the cluster center:

```
for ( (k, v) <- clusterAssignments.toSeq.sortBy(_._1)) {
  println(s"Cluster $k:")
  val m = v.toSeq.sortBy(_._5)
  println(m.take(20).map { case (_, title, genres, _, d) =>
    (title, genres, d) }.mkString("\n"))
  println("=====\n")
}
```

The following screenshot is an example output. Note that your output might differ due to random initializations of both the recommendation and clustering model.

```
Cluster 0:
(Last Time I Saw Paris, The (1954),Drama,0.27390666869786695)
(Quiz Show (1994),Drama,0.4747831636277422)
(Vertigo (1958),Mystery Thriller,0.48534208687692343)
(Spellbound (1945),Mystery Romance Thriller,0.4926221112685535)
(Casablanca (1942),Drama Romance War,0.49940194962368567)
(African Queen, The (1951),Action Adventure Romance War,0.5187502052689528)
(Amadeus (1984),Drama Mystery,0.5272552880790345)
(Farewell to Arms, A (1932),Romance War,0.5363608755281067)
(Cat on a Hot Tin Roof (1958),Drama,0.5497562196607095)
(Third Man, The (1949),Mystery Thriller,0.5497731051647746)
(Dial M for Murder (1954),Mystery Thriller,0.5622477772149612)
(North by Northwest (1959),Comedy Thriller,0.5702331060033082)
(20,000 Leagues Under the Sea (1954),Adventure Children's Fantasy Sci-Fi,0.5881687768024192)
(Right Stuff, The (1983),Drama,0.6002418388739418)
(Rear Window (1954),Mystery Thriller,0.6232262641317354)
(Manchurian Candidate, The (1962),Film-Noir Thriller,0.6233301146337812)
(Substance of Fire, The (1996),Drama,0.6252591340497877)
(M*A*S*H (1970),Comedy War,0.63105245443614)
(Butch Cassidy and the Sundance Kid (1969),Action Comedy Western,0.6337504848523161)
(Blue Angel, The (Blaue Engel, Der) (1930),Drama,0.6342821363539322)
```

The first cluster

The first cluster, labeled 0, seems to contain a lot of old movies from the 1940s, 1950s, and 1960s, as well as a scattering of recent dramas.

```
Cluster 1:
(Amityville 1992: It's About Time (1992),Horror,0.1478043405622148)
(Amityville: A New Generation (1993),Horror,0.1478043405622148)
(Gordy (1995),Comedy,0.15051585838791465)
(Machine, The (1994),Comedy Horror,0.176865932564681)
(Amityville: Dollhouse (1996),Horror,0.17898379655862778)
(Venice/Venice (1992),Drama,0.19738131555708463)
(Somebody to Love (1994),Drama,0.2278813718368857)
(Boys in Venice (1996),Drama,0.2278813718368857)
(Falling in Love Again (1980),Comedy,0.2340143978726976)
(3 Ninjas: High Noon At Mega Mountain (1998),Action Children's,0.23903016507829816)
(Babyfever (1994),Comedy Drama,0.24176557927323153)
(Beyond Bedlam (1993),Drama Horror,0.2489480589001102)
(Getting Away With Murder (1996),Comedy,0.2530960279675358)
(Police Story 4: Project S (Chao ji ji hua) (1993),Action,0.25942902404443574)
(Mighty, The (1998),Drama,0.27817019934466347)
(Johnny 100 Pesos (1993),Action Drama,0.2870737627453892)
(King of New York (1990),Action Crime,0.28853211361643927)
(Further Gesture, A (1996),Drama,0.29378208871990685)
(Shadow of Angels (Schatten der Engel) (1976),Drama,0.29529253258337934)
(Homage (1995),Drama,0.29529253258337934)
```

The second cluster

The second cluster has a few horror movies in a prominent position, while the rest of the movies are less clear, but dramas are common too.

```
Cluster 2:
(House Party 3 (1994),Comedy,0.5792798401193011)
(Cops and Robbersons (1994),Comedy,0.6121886776465748)
(Pagemaster, The (1994),Action Adventure Animation Children's Fantasy,0.6126925309798513)
(Fausto (1993),Comedy,0.6220018406977679)
(Stag (1997),Action Thriller,0.6694984978987776)
(Ill Gotten Gains (1997),Drama,0.7021111594974133)
(All Things Fair (1996),Drama,0.7365539555740591)
(Day the Sun Turned Cold, The (Tianguo niezi) (1994),Drama,0.7447955673545115)
(Chasers (1994),Comedy,0.7459052286323937)
(Pyromaniac's Love Story, A (1995),Comedy Romance,0.7746300046654674)
(Robocop 3 (1993),Sci-Fi Thriller,0.8075493355683138)
(American Strays (1996),Action,0.8375011873201667)
(Scout, The (1994),Drama,0.8455857296456323)
(Metro (1997),Action,0.8488282233075414)
(Sunchaser, The (1996),Drama,0.8855757549882701)
(Across the Sea of Time (1995),Documentary,0.9132140236347115)
(Big Bully (1996),Comedy Drama,0.9134404160863872)
(Wife, The (1995),Comedy Drama,0.9136501322150961)
(Big Squeeze, The (1996),Comedy Drama,0.9191497196405036)
(Shooter, The (1995),Action,0.9309878751600442)
```

The third cluster

The third cluster is not clear-cut but has a fair number of comedy and drama movies.

```
Cluster 3:
(King of the Hill (1993),Drama,0.27977910057590455)
(Love and Other Catastrophes (1996),Romance,0.5616301951805126)
(All Over Me (1997),Drama,0.5827486944870316)
(Scream of Stone (Schrei aus Stein) (1991),Drama,0.5990653123876859)
(Witness (1985),Drama Romance Thriller,0.6251178451970778)
(I Can't Sleep (J'ai pas sommeil) (1994),Drama Thriller,0.6810378136145686)
(Ed's Next Move (1996),Comedy,0.6821637177989938)
(Suture (1993),Film-Noir Thriller,0.7247521033315935)
(Sex, Lies, and Videotape (1989),Drama,0.7431922597566741)
(Double Happiness (1994),Drama,0.770636268189707)
(Wild Bill (1995),Western,0.7860403052412567)
(Smoke (1995),Drama,0.7929521364994968)
(Lover's Knot (1996),Comedy,0.7952419475534458)
(Howling, The (1981),Comedy Horror,0.7958806811974748)
(Price Above Rubies, A (1998),Drama,0.797523480324549)
(Wooden Man's Bride, The (Wu Kui) (1994),Drama,0.8035270945874013)
(Nelly & Monsieur Arnaud (1995),Drama,0.8050334619603677)
(Gate of Heavenly Peace, The (1995),Documentary,0.807333841007159)
(Substance of Fire, The (1996),Drama,0.8143443692443669)
(Grifters, The (1990),Crime Drama Film-Noir,0.8234461534563621)
```

The fourth cluster

The next cluster is more clearly associated with dramas and contains some foreign language films in particular.

```
Cluster 4:
(Outbreak (1995),Action Drama Thriller,0.4526691989349761)
(River Wild, The (1994),Action Thriller,0.46017763132846606)
(Moonlight and Valentino (1995),Drama Romance,0.472253677017327)
(Blue Chips (1994),Drama,0.5103978205046279)
(Outlaw, The (1943),Western,0.5346838076035247)
(Air Up There, The (1994),Comedy,0.5721399113559971)
(Touch (1997),Romance,0.5873709976348385)
(Private Benjamin (1980),Comedy,0.5915397936710273)
(Angela (1995),Drama,0.6075617445146397)
(Sword in the Stone, The (1963),Animation Children's,0.6165719141792315)
(Mr. Wonderful (1993),Comedy Romance,0.6181379459010301)
(Maverick (1994),Action Comedy Western,0.6316402376687157)
(Cool Runnings (1993),Comedy,0.6462611091600288)
(Courage Under Fire (1996),Drama War,0.6603376056624485)
(I.Q. (1994),Comedy Romance,0.66691874141152)
(Ransom (1996),Drama Thriller,0.6755383826704695)
(City of Angels (1998),Romance,0.6756718112001122)
(Firm, The (1993),Drama Thriller,0.6769576000019328)
(Santa Clause, The (1994),Children's Comedy,0.6795328449586006)
(Cliffhanger (1993),Action Adventure Crime,0.703261186148323)
```

The last cluster

The final cluster seems to be related predominantly to action and thrillers as well as romance movies, and seems to contain a number of relatively popular movies.

As you can see, it is not always straightforward to determine exactly what each cluster represents. However, there is some evidence here that the clustering is picking out attributes or commonalities between groups of movies, which might not be immediately obvious based only on the movie titles and genres (such as a foreign language segment, a classic movie segment, and so on). If we had more metadata available, such as directors, actors, and so on, we might find out more details about the defining features of each cluster.

We leave it as an exercise for you to perform a similar investigation into the clustering of the user factors. We have already created the input vectors in the userVectors variable, so you can train a K-means model on these vectors. After that, in order to evaluate the clusters, you would need to investigate the closest users for each cluster center (as we did for movies) and see if some common characteristics can be identified from the movies they have rated or the user metadata available.

Evaluating the performance of clustering models

Like models such as regression, classification, and recommendation engines, there are many evaluation metrics that can be applied to clustering models to analyze their performance and the goodness of the clustering of the data points. Clustering evaluation is generally divided into either internal or external evaluation. Internal evaluation refers to the case where the same data used to train the model is used for evaluation. External evaluation refers to using data external to the training data for evaluation purposes.

Internal evaluation metrics

Common internal evaluation metrics include the WCSS we covered earlier (which is exactly the K-means objective function), the Davies-Bouldin index, the Dunn Index, and the silhouette coefficient. All these measures tend to reward clusters where elements within a cluster are relatively close together, while elements in different clusters are relatively far away from each other.

> The Wikipedia page on clustering evaluation at http://en.wikipedia.org/wiki/Cluster_analysis#Internal_evaluation has more details.

External evaluation metrics

Since clustering can be thought of as unsupervised classification, if we have some form of labeled (or partially labeled) data available, we could use these labels to evaluate a clustering model. We can make predictions of clusters (that is, the class labels) using the model and evaluate the predictions against the true labels using metrics similar to some that we saw for classification evaluation (that is, based on true positive and negative and false positive and negative rates).

These include the Rand measure, F-measure, Jaccard index, and others.

> See http://en.wikipedia.org/wiki/Cluster_analysis#External_evaluation for more information on external evaluation for clustering.

Computing performance metrics on the MovieLens dataset

MLlib provides a convenient `computeCost` function to compute the WCSS objective function given a RDD [`Vector`]. We will compute this metric for the following movie and user training data:

```
val movieCost = movieClusterModel.computeCost(movieVectors)
val userCost = userClusterModel.computeCost(userVectors)
println("WCSS for movies: " + movieCost)
println("WCSS for users: " + userCost)
```

This should output the result similar to the following one:

```
WCSS for movies: 2586.0777166339426
WCSS for users: 1403.4137493396831
```

Tuning parameters for clustering models

In contrast to many of the other models we have come across so far, K-means only has one parameter that can be tuned. This is K, the number of cluster centers chosen.

Selecting K through cross-validation

As we have done with classification and regression models, we can apply cross-validation techniques to select the optimal number of clusters for our model. This works in much the same way as for supervised learning methods. We will split the dataset into a training set and a test set. We will then train a model on the training set and compute the evaluation metric of interest on the test set.

We will do this for the movie clustering using the built-in WCSS evaluation metric provided by MLlib in the following code, using a 60 percent / 40 percent split between the training set and test set:

```
val trainTestSplitMovies = movieVectors.randomSplit(Array(0.6, 0.4), 123)
val trainMovies = trainTestSplitMovies(0)
val testMovies = trainTestSplitMovies(1)
val costsMovies = Seq(2, 3, 4, 5, 10, 20).map { k => (k, KMeans.train(trainMovies, numIterations, k, numRuns).computeCost(testMovies))
}
println("Movie clustering cross-validation:")
costsMovies.foreach { case (k, cost) => println(f"WCSS for K=$k id $cost%2.2f") }
```

This should give results that look something like the ones shown here.

The output of movie clustering cross-validation is:

```
Movie clustering cross-validation
WCSS for K=2 id 942.06
WCSS for K=3 id 942.67
WCSS for K=4 id 950.35
WCSS for K=5 id 948.20
WCSS for K=10 id 943.26
WCSS for K=20 id 947.10
```

We can observe that the WCSS decreases as the number of clusters increases, up to a point. It then begins to increase. Another common pattern observed in the WCSS in cross-validation for K-means is that the metric continues to decrease as K increases, but at a certain point, the rate of decrease flattens out substantially. The value of K at which this occurs is generally selected as the optimal K parameter (this is sometimes called the elbow point, as this is where the line kinks when drawn as a graph).

In our case, we might select a value of 10 for K, based on the preceding results. Also, note that the clusters that are computed by the model are often used for purposes that require some human interpretation (such as the cases of movie and customer segmentation we mentioned earlier). Therefore, this consideration also impacts the choice of K, as although a higher value of K might be more optimal from the mathematical point of view, it might be more difficult to understand and interpret many clusters.

For completeness, we will also compute the cross-validation metrics for user clustering:

```
val trainTestSplitUsers = userVectors.randomSplit(Array(0.6, 0.4), 123)
val trainUsers = trainTestSplitUsers(0)
val testUsers = trainTestSplitUsers(1)
val costsUsers = Seq(2, 3, 4, 5, 10, 20).map { k => (k,
KMeans.train(trainUsers, numIterations, k,
numRuns).computeCost(testUsers)) }
println("User clustering cross-validation:")
costsUsers.foreach { case (k, cost) => println(f"WCSS for K=$k id
$cost%2.2f") }
```

We will see a pattern that is similar to the movie case:

```
User clustering cross-validation:
WCSS for K=2 id 544.02
WCSS for K=3 id 542.18
WCSS for K=4 id 542.38
WCSS for K=5 id 542.33
WCSS for K=10 id 539.68
WCSS for K=20 id 541.21
```

> Note that your results may differ slightly due to random initialization of the clustering models.

Summary

In this chapter, we explored a new class of model that learns structure from unlabeled data—unsupervised learning. We worked through required input data, feature extraction, and saw how to use the output of one model (a recommendation model in our example) as the input to another model (our K-means clustering model). Finally, we evaluated the performance of the clustering model, both using manual interpretation of the cluster assignments and using mathematical performance metrics.

In the next chapter, we will cover another type of unsupervised learning used to reduce our data down to its most important features or components—dimensionality reduction models.

8
Dimensionality Reduction with Spark

Over the course of this chapter, we will continue our exploration of unsupervised learning models in the form of **dimensionality reduction**.

Unlike the models we have covered so far, such as regression, classification, and clustering, dimensionality reduction does not focus on making predictions. Instead, it tries to take a set of input data with a feature dimension D (that is, the length of our feature vector) and extract a representation of the data of dimension k, where k is usually significantly smaller than D. It is, therefore, a form of preprocessing or feature transformation rather than a predictive model in its own right.

It is important that the representation that is extracted should still be able to capture a large proportion of the variability or structure of the original data. The idea behind this is that most data sources will contain some form of underlying structure. This structure is typically unknown (often called latent features or latent factors), but if we can uncover some of this structure, our models could learn this structure and make predictions from it rather than from the data in its raw form, which might be noisy or contain many irrelevant features. In other words, dimensionality reduction throws away some of the noise in the data and keeps the hidden structure that is present.

In some cases, the dimensionality of the raw data is far higher than the number of data points we have, so without dimensionality reduction, it would be difficult for other machine learning models, such as classification and regression, to learn anything, as they need to fit a number of parameters that is far larger than the number of training examples (in this sense, these methods bear some similarity to the regularization approaches that we have seen used in classification and regression).

A few use cases of dimensionality reduction techniques include:

- Exploratory data analysis
- Extracting features to train other machine learning models
- Reducing storage and computation requirements for very large models in the prediction phase (for example, a production system that makes predictions)
- Reducing a large group of text documents down to a set of hidden topics or concepts
- Making learning and generalization of models easier when our data has a very large number of features (for example, when working with text, sound, images, or video data, which tends to be very high-dimensional)

In this chapter, we will:

- Introduce the types of dimensionality reduction models available in MLlib
- Work with images of faces to extract features suitable for dimensionality reduction
- Train a dimensionality reduction model using MLlib
- Visualize and evaluate the results
- Perform parameter selection for our dimensionality reduction model

Types of dimensionality reduction

MLlib provides two models for dimensionality reduction; these models are closely related to each other. They are **Principal Components Analysis (PCA)** and **Singular Value Decomposition (SVD)**.

Principal Components Analysis

PCA operates on a data matrix X and seeks to extract a set of k principal components from X. The principal components are each uncorrelated to each other and are computed such that the first principal component accounts for the largest variation in the input data. Each subsequent principal component is, in turn, computed such that it accounts for the largest variation, provided that it is independent of the principal components computed so far.

In this way, the k principal components returned are guaranteed to account for the highest amount of variation in the input data possible. Each principal component, in fact, has the same feature dimensionality as the original data matrix. Hence, a projection step is required in order to actually perform dimensionality reduction, where the original data is projected into the *k-dimensional* space represented by the principal components.

Singular Value Decomposition

SVD seeks to decompose a matrix X of dimension *m x n* into three component matrices:

- U of dimension *m x m*
- S, a diagonal matrix of size *m x n*; the entries of S are referred to as the **singular values**
- V^T of dimension *n x n*

$$X = U * S * V^T$$

Looking at the preceding formula, it appears that we have not reduced the dimensionality of the problem at all, as by multiplying U, S, and V, we reconstruct the original matrix. In practice, the truncated SVD is usually computed. That is, only the top k singular values, which represent the most variation in the data, are kept, while the rest are discarded. The formula to reconstruct X based on the component matrices is then approximate:

$$X \sim U_k * S_k * V_{k\,T}$$

An illustration of the truncated SVD is shown here:

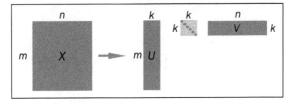

The truncated Singular Value Decomposition

Keeping the top k singular values is similar to keeping the top k principal components in PCA. In fact, SVD and PCA are directly related, as we will see a little later in this chapter.

> A detailed mathematical treatment of both PCA and SVD is beyond the scope of this book.
>
> An overview of dimensionality reduction can be found in the Spark documentation at http://spark.apache.org/docs/latest/mllib-dimensionality-reduction.html.
>
> The following links contain a more in-depth mathematical overview of PCA and SVD, respectively: http://en.wikipedia.org/wiki/Principal_component_analysis and http://en.wikipedia.org/wiki/Singular_value_decomposition.

Relationship with matrix factorization

PCA and SVD are both matrix factorization techniques, in the sense that they decompose a data matrix into subcomponent matrices, each of which has a lower dimension (or rank) than the original matrix. Many other dimensionality reduction techniques are based on matrix factorization.

You might remember another example of matrix factorization, that is, collaborative filtering, that we have already seen in *Chapter 4, Building a Recommendation Engine with Spark*. Matrix factorization approaches to collaborative filtering work by factorizing the ratings matrix into two components: the user factor matrix and the item factor matrix. Each of these has a lower dimension than the original data, so these methods also act as dimensionality reduction models.

> Many of the best performing approaches to collaborative filtering include models based on SVD. Simon Funk's approach to the Netflix prize is a famous example. You can look at it at http://sifter.org/~simon/journal/20061211.html.

Clustering as dimensionality reduction

The clustering models we covered in the previous chapter can also be used for a form of dimensionality reduction. This works in the following way:

- Assume that we cluster our high-dimensional feature vectors using a K-means clustering model, with k clusters. The result is a set of k cluster centers.

- We can represent each of our original data points in terms of how far it is from each of these cluster centers. That is, we can compute the distance of a data point to each cluster center. The result is a set of k distances for each data point.

- These *k* distances can form a new vector of dimension *k*. We can now represent our original data as a new vector of lower dimension, relative to the original feature dimension.

Depending on the distance metric used, this can result in both dimensionality reduction and a form of nonlinear transformation of the data, allowing us to learn a more complex model while still benefiting from the speed and scalability of a linear model. For example, using a Gaussian or exponential distance function can approximate a very complex nonlinear feature transformation.

Extracting the right features from your data

As with all machine learning models we have explored so far, dimensionality reduction models also operate on a feature vector representation of our data.

For this chapter, we will dive into the world of image processing, using the **Labeled Faces in the Wild (LFW)** dataset of facial images. This dataset contains over 13,000 images of faces generally taken from the Internet and belonging to well-known public figures. The faces are labeled with the person's name.

Extracting features from the LFW dataset

In order to avoid having to download and process a very large dataset, we will work with a subset of the images, using people who have names that start with an "A". This dataset can be downloaded from http://vis-www.cs.umass.edu/lfw/lfw-a.tgz.

> For more details and other variants of the data, visit http://vis-www.cs.umass.edu/lfw/.
>
> The original research paper reference is:
>
> *Gary B. Huang, Manu Ramesh, Tamara Berg,* and *Erik Learned-Miller. Labeled Faces in the Wild: A Database for Studying Face Recognition in Unconstrained Environments.* University of Massachusetts, Amherst, Technical Report 07-49, October, 2007.
>
> It can be downloaded from http://vis-www.cs.umass.edu/lfw/lfw.pdf.

Unzip the data using the following command:

```
>tar xfvz lfw-a.tgz
```

This will create a folder called lfw, which contains a number of subfolders, one for each person.

Exploring the face data

Start up your Spark Scala console by ensuring that you allocate sufficient memory, as dimensionality reduction methods can be quite computationally expensive:

```
>./SPARK_HOME/bin/spark-shell --driver-memory 2g
```

Now that we've unzipped the data, we face a small challenge. Spark provides us with a way to read text files and custom Hadoop input data sources. However, there is no built-in functionality to allow us to read images.

Spark provides a method called wholeTextFiles, which allows us to operate on entire files at once, compared to the textFile method that we have been using so far, which operates on the individual lines within a text file (or multiple files).

We will use the wholeTextFiles method to access the location of each file. Using these file paths, we will write custom code to load and process the images. In the following example code, we will use PATH to refer to the directory in which you extracted the lfw subdirectory.

We can use a wildcard path specification (using the * character highlighted in the following code snippet) to tell Spark to look in each directory under the lfw directory for files:

```
val path = "/PATH/lfw/*"
val rdd = sc.wholeTextFiles(path)
val first = rdd.first
println(first)
```

Running the first command might take a little time, as Spark first scans the specified directory structure for all available files. Once completed, you should see output similar to the one shown here:

```
first: (String, String) =  (file:/PATH/lfw/Aaron_Eckhart
/Aaron_Eckhart_0001.jpg,░░░░??JFIF????? ...
```

You will see that `wholeTextFiles` returns an RDD that contains key-value pairs, where the key is the file location while the value is the content of the entire text file. For our purposes, we only care about the file path, as we cannot work directly with the image data as a string (notice that it is displayed as "binary nonsense" in the shell output).

Let's extract the file paths from the RDD. Note that earlier, the file path starts with the `file:` text. This is used by Spark when reading files in order to differentiate between different filesystems (for example, `file://` for the local filesystem, `hdfs://` for HDFS, `s3n://` for Amazon S3, and so on).

In our case, we will be using custom code to read the images, so we don't need this part of the path. Thus, we will remove it with the following `map` function:

```
val files = rdd.map { case (fileName, content) =>
fileName.replace("file:", "") }
println(files.first)
```

This should display the file location with the `file:` prefix removed:

`/PATH/lfw/Aaron_Eckhart/Aaron_Eckhart_0001.jpg`

Next, we will see how many files we are dealing with:

```
println(files.count)
```

Running these commands creates a lot of noisy output in the Spark shell, as it outputs all the file paths that are read to the console. Ignore this part, but after the command has completed, the output should look something like this:

```
..., /PATH/lfw/Azra_Akin/Azra_Akin_0003.jpg:0+19927,
/PATH/lfw/Azra_Akin/Azra_Akin_0004.jpg:0+16030

...

14/09/18 20:36:25 INFO SparkContext: Job finished: count at
<console>:19, took 1.151955 s
```
1055

So, we can see that we have 1055 images to work with.

Visualizing the face data

Although there are a few tools available in Scala or Java to display images, this is one area where Python and the matplotlib library shine. We will use Scala to process and extract the images and run our models and IPython to display the actual images.

You can run a separate IPython Notebook by opening a new terminal window and launching a new notebook:

`>ipython notebook`

 Note that if using Python Notebook, you should first execute the following code snippet to ensure that the images are displayed inline after each notebook cell (including the % character): `%pylab inline`.

Alternatively, you can launch a plain IPython console without the web notebook, enabling the `pylab` plotting functionality using the following command:

`>ipython --pylab`

The dimensionality reduction techniques in MLlib are only available in Scala or Java at the time of writing this book, so we will continue to use the Scala Spark shell to run the models. Therefore, you won't need to run a PySpark console.

 We have provided the full Python code with this chapter as a Python script as well as in the IPython Notebook format. For instructions on installing IPython, see the code bundle.

Let's display the image given by the first path we extracted earlier using matplotlib's `imread` and `imshow` functions:

```
path = "/PATH/lfw/PATH/lfw/Aaron_Eckhart/Aaron_Eckhart_0001.jpg"
ae = imread(path)
imshow(ae)
```

 You should see the image displayed in your Notebook (or in a pop-up window if you are using the standard IPython shell). Note that we have not shown the image here.

Extracting facial images as vectors

While a full treatment of image processing is beyond the scope of this book, we will need to know a few basics to proceed. Each color image can be represented as a three-dimensional array, or matrix, of pixels. The first two dimensions, that is the *x* and *y* axes, represent the position of each pixel, while the third dimension represents the **red, blue, and green (RGB)** color values for each pixel.

A grayscale image only requires one value per pixel (there are no RGB values), so it can be represented as a plain two-dimensional matrix. For many image-processing and machine learning tasks related to images, it is common to operate on grayscale images. We will do this here by converting the color images to grayscale first.

It is also a common practice in machine learning tasks to represent an image as a vector, instead of a matrix. We do this by concatenating each row (or alternatively, each column) of the matrix together to form a long vector (this is known as **reshaping**). In this way, each raw, grayscale image matrix is transformed into a feature vector that is usable as input to a machine learning model.

Fortunately for us, the built-in Java **Abstract Window Toolkit** (**AWT**) contains various basic image-processing functions. We will define a few utility functions to perform this processing using the `java.awt` classes.

Loading images

The first of these is a function to read an image from a file:

```
import java.awt.image.BufferedImage
def loadImageFromFile(path: String): BufferedImage = {
  import javax.imageio.ImageIO
  import java.io.File
  ImageIO.read(new File(path))
}
```

This returns an instance of a `java.awt.image.BufferedImage` class, which stores the image data and provides a number of useful methods. Let's test it out by loading the first image into our Spark shell:

```
val aePath = "/PATH/lfw/Aaron_Eckhart/Aaron_Eckhart_0001.jpg"
val aeImage = loadImageFromFile(aePath)
```

You should see the image details displayed in the shell:

```
aeImage: java.awt.image.BufferedImage = BufferedImage@f41266e: type =
5 ColorModel: #pixelBits = 24 numComponents = 3 color space =
java.awt.color.ICC_ColorSpace@7e420794 transparency = 1 has alpha =
false isAlphaPre = false ByteInterleavedRaster: width = 250 height =
250 #numDataElements 3 dataOff[0] = 2
```

There is quite a lot of information here. Of particular interest to us is that the image width and height are 250 pixels, and as we can see, there are three components (that is, the RGB values) that are highlighted in the preceding output.

Converting to grayscale and resizing the images

The next function we will define will take the image that we have loaded with our preceding function, convert the image from color to grayscale, and resize the image's width and height.

These steps are not strictly necessary, but both steps are done in many cases for efficiency purposes. Using RGB color images instead of grayscale increases the amount of data to be processed by a factor of 3. Similarly, larger images increase the processing and storage overhead significantly. Our raw 250 x 250 images represent 187,500 data points per image using three color components. For a set of 1055 images, this is 197,812,500 data points. Even if stored as integer values, each value stored takes 4 bytes of memory, so just 1055 images represent around 800 MB of memory! As you can see, image-processing tasks can quickly become extremely memory intensive.

If we convert to grayscale and resize the images to, say, 50 x 50 pixels, we only require 2500 data points per image. For our 1055 images, this equates to 10 MB of memory, which is far more manageable for illustrative purposes.

 Another reason to resize is that MLlib's PCA model works best on *tall and skinny* matrices with less than 10,000 columns. We will have 2500 columns (that is, each pixel becomes an entry in our feature vector), so we will come in well below this restriction.

Let's define our processing function. We will do the grayscale conversion and resizing in one step, using the `java.awt.image` package:

```
def processImage(image: BufferedImage, width: Int, height: Int):
BufferedImage = {
  val bwImage = new BufferedImage
    (width, height, BufferedImage.TYPE_BYTE_GRAY)
```

```
    val g = bwImage.getGraphics()
    g.drawImage(image, 0, 0, width, height, null)
    g.dispose()
    bwImage
}
```

The first line of the function creates a new image of the desired width and height and specifies a grayscale color model. The third line draws the original image onto this newly created image. The `drawImage` method takes care of the color conversion and resizing for us! Finally, we return the new, processed image.

Let's test this out on our sample image. We will convert it to grayscale and resize it to 100 x 100 pixels:

```
    val grayImage = processImage(aeImage, 100, 100)
```

You should see the following output in the console:

```
grayImage: java.awt.image.BufferedImage = BufferedImage@21f8ea3b:
type = 10 ColorModel: #pixelBits = 8 numComponents = 1 color space =
java.awt.color.ICC_ColorSpace@5cd9d8e9 transparency = 1 has alpha =
false isAlphaPre = false ByteInterleavedRaster: width = 100 height =
100 #numDataElements 1 dataOff[0] = 0
```

As you can see from the highlighted output, the image's width and height are indeed 100, and the number of color components is 1.

Next, we will save the processed image to a temporary location so that we can read it back and display it in our IPython console:

```
    import javax.imageio.ImageIO
    import java.io.File
    ImageIO.write(grayImage, "jpg", new File("/tmp/aeGray.jpg"))
```

You should see a result of `true` displayed in your console, indicating that we successfully saved the image to the `aeGray.jpg` file in our `/tmp` directory.

Finally, we will read the image in Python and use matplotlib to display the image. Type the following code into your IPython Notebook or shell (remember that this should be open in a new terminal window):

```
    tmpPath = "/tmp/aeGray.jpg"
    aeGary = imread(tmpPath)
    imshow(aeGary, cmap=plt.cm.gray)
```

This should display the image (note again, we haven't shown the image here). You should see that it is grayscale and of slightly worse quality as compared to the original image. Furthermore, you will notice that the scale of the axes are different, representing the new 100 x 100 dimension instead of the original 250 x 250 size.

Extracting feature vectors

The final step in the processing pipeline is to extract the actual feature vectors that will be the input to our dimensionality reduction model. As we mentioned earlier, the raw grayscale pixel data will be our features. We will form the vectors by flattening out the two-dimensional pixel matrix. The `BufferedImage` class provides a utility method to do just this, which we will use in our function:

```
def getPixelsFromImage(image: BufferedImage): Array[Double] = {
  val width = image.getWidth
  val height = image.getHeight
  val pixels = Array.ofDim[Double](width * height)
  image.getData.getPixels(0, 0, width, height, pixels)
}
```

We can then combine these three functions into one utility function that takes a file location together with the desired image's width and height and returns the raw `Array[Double]` value that contains the pixel data:

```
def extractPixels(path: String, width: Int, height: Int):
Array[Double] = {
  val raw = loadImageFromFile(path)
  val processed = processImage(raw, width, height)
  getPixelsFromImage(processed)
}
```

Applying this function to each element of the RDD that contains all the image file paths will give us a new RDD that contains the pixel data for each image. Let's do this and inspect the first few elements:

```
val pixels = files.map(f => extractPixels(f, 50, 50))
println(pixels.take(10).map(_.take(10).mkString
("", ",", ", ...")).mkString("\n"))
```

You should see output similar to this:

```
0.0,0.0,0.0,0.0,0.0,0.0,1.0,1.0,0.0,0.0, ...
241.0,243.0,245.0,244.0,231.0,205.0,177.0,160.0,150.0,147.0, ...
253.0,253.0,253.0,253.0,253.0,253.0,254.0,254.0,253.0,253.0, ...
```

```
244.0,244.0,243.0,242.0,241.0,240.0,239.0,239.0,237.0,236.0, ...
44.0,47.0,47.0,49.0,62.0,116.0,173.0,223.0,232.0,233.0, ...
0.0,0.0,0.0,0.0,0.0,0.0,0.0,0.0,0.0,0.0, ...
1.0,1.0,1.0,1.0,1.0,1.0,1.0,1.0,0.0,0.0, ...
26.0,26.0,27.0,26.0,24.0,24.0,25.0,26.0,27.0,27.0, ...
240.0,240.0,240.0,240.0,240.0,240.0,240.0,240.0,240.0,240.0, ...
0.0,0.0,0.0,0.0,0.0,0.0,0.0,0.0,0.0,0.0, ...
```

The final step is to create an MLlib `Vector` instance for each image. We will cache the RDD to speed up our later computations:

```
import org.apache.spark.mllib.linalg.Vectors
val vectors = pixels.map(p => Vectors.dense(p))
vectors.setName("image-vectors")
vectors.cache
```

> We used the `setName` function earlier to assign an RDD a name. In this case, we called it `image-vectors`. This is so that we can later identify it more easily when looking at the Spark web interface.

Normalization

It is a common practice to standardize input data prior to running dimensionality reduction models, in particular for PCA. As we did in *Chapter 5, Building a Classification Model with Spark*, we will do this using the built-in `StandardScaler` provided by MLlib's `feature` package. We will only subtract the mean from the data in this case:

```
import org.apache.spark.mllib.linalg.Matrix
import org.apache.spark.mllib.linalg.distributed.RowMatrix
import org.apache.spark.mllib.feature.StandardScaler
val scaler = new StandardScaler
(withMean = true, withStd = false).fit(vectors)
```

Calling `fit` triggers a computation on our `RDD[Vector]`. You should see output similar to the one shown here:

```
...
14/09/21 11:46:58 INFO SparkContext: Job finished: reduce at
RDDFunctions.scala:111, took 0.495859 s
scaler: org.apache.spark.mllib.feature.StandardScalerModel =
org.apache.spark.mllib.feature.StandardScalerModel@6bb1a1a1
```

Dimensionality Reduction with Spark

> Note that subtracting the mean works for dense input data. However, for sparse vectors, subtracting the mean vector from each input will transform the sparse data into dense data. For very high-dimensional input, this will likely exhaust the available memory resources, so it is not advisable.

Finally, we will use the returned `scaler` to transform the raw image vectors to vectors with the column means subtracted:

```
val scaledVectors = vectors.map(v => scaler.transform(v))
```

We mentioned earlier that the resized grayscale images would take up around 10 MB of memory. Indeed, you can take a look at the memory usage in the Spark application monitor storage page by going to `http://localhost:4040/storage/` in your web browser.

Since we gave our RDD of image vectors a friendly name of `image-vectors`, you should see something like the following screenshot (note that as we are using `Vector[Double]`, each element takes up 8 bytes instead of 4 bytes; hence, we actually use 20 MB of memory):

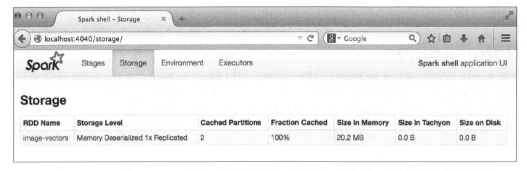

Size of image vectors in memory

Training a dimensionality reduction model

Dimensionality reduction models in MLlib require vectors as inputs. However, unlike clustering that operated on an `RDD[Vector]`, PCA and SVD computations are provided as methods on a distributed `RowMatrix` (this difference is largely down to syntax, as a `RowMatrix` is simply a wrapper around an `RDD[Vector]`).

Running PCA on the LFW dataset

Now that we have extracted our image pixel data into vectors, we can instantiate a new `RowMatrix` and call the `computePrincipalComponents` method to compute the top K principal components of our distributed matrix:

```
import org.apache.spark.mllib.linalg.Matrix
import org.apache.spark.mllib.linalg.distributed.RowMatrix
val matrix = new RowMatrix(scaledVectors)
val K = 10
val pc = matrix.computePrincipalComponents(K)
```

You will likely see quite a lot of output in your console while the model runs.

> If you see warnings such as **WARN LAPACK: Failed to load implementation from: com.github.fommil.netlib. NativeSystemLAPACK** or **WARN LAPACK: Failed to load implementation from: com.github.fommil.netlib. NativeRefLAPACK**, you can safely ignore these.
>
> This means that the underlying linear algebra libraries used by MLlib could not load native routines. In this case, a Java-based fallback will be used, which is slower, but there is nothing to worry about for the purposes of this example.

Once the model training is complete, you should see a result displayed in the console that looks similar to the following one:

```
pc: org.apache.spark.mllib.linalg.Matrix =
-0.023183157256614906   -0.010622723054037303   ... (10 total)
-0.023960537953442107   -0.011495966728461177   ...
-0.024397470862198022   -0.013512219690177352   ...
-0.02463158818330343    -0.014758658113862178   ...
-0.024941633606137027   -0.014878858729655142   ...
-0.02525998879466241    -0.014602750644394844   ...
-0.025494722450369593   -0.014678013626511024   ...
-0.02604194423255582    -0.01439561589951032    ...
-0.025942214214865228   -0.013907665261197633   ...
-0.026151551334429365   -0.014707035797934148   ...
-0.026106572186134578   -0.016701471378568943   ...
```

```
-0.026242986173995755    -0.016254664123732318   ...
-0.02573628754284022     -0.017185663918352894   ...
-0.02545319635905169     -0.01653357295561698    ...
-0.025325893980995124    -0.0157082218373399...
```

Visualizing the Eigenfaces

Now that we have trained our PCA model, what is the result? Let's inspect the dimensions of the resulting matrix:

```
val rows = pc.numRows
val cols = pc.numCols
println(rows, cols)
```

As you should see from your console output, the matrix of principal components has 2500 rows and 10 columns:

```
(2500,10)
```

Recall that the dimension of each image is 50 x 50, so here, we have the top 10 principal components, each with a dimension identical to that of the input images. These principal components can be thought of as the set of latent (or hidden) features that capture the greatest variation in the original data.

> In facial recognition and image processing, these principal components are often referred to as **Eigenfaces**, as PCA is closely related to the eigenvalue decomposition of the covariance matrix of the original data.
> See http://en.wikipedia.org/wiki/Eigenface for more details.

Since each principal component is of the same dimension as the original images, each component can itself be thought of and represented as an image, making it possible to visualize the Eigenfaces as we would the input images.

As we have often done in this book, we will use functionality from the Breeze linear algebra library as well as Python's numpy and matplotlib to visualize the Eigenfaces.

First, we will extract the pc variable (an MLlib matrix) into a Breeze DenseMatrix:

```
import breeze.linalg.DenseMatrix

val pcBreeze = new DenseMatrix(rows, cols, pc.toArray)
```

Breeze provides a useful function within the linalg package to write the matrix out as a CSV file. We will use this to save the principal components to a temporary CSV file:

```
import breeze.linalg.csvwrite
csvwrite(new File("/tmp/pc.csv"), pcBreeze)
```

Next, we will load the matrix in IPython and visualize the principal components as images. Fortunately, numpy provides a utility function to read the matrix from the CSV file we created:

```
pcs = np.loadtxt("/tmp/pc.csv", delimiter=",")
print(pcs.shape)
```

You should see the following output, confirming that the matrix we read has the same dimensions as the one we saved:

(2500, 10)

We will need a utility function to display the images, which we define here:

```
def plot_gallery(images, h, w, n_row=2, n_col=5):
    """Helper function to plot a gallery of portraits"""
    plt.figure(figsize=(1.8 * n_col, 2.4 * n_row))
    plt.subplots_adjust(bottom=0, left=.01, right=.99, top=.90, hspace=.35)
    for i in range(n_row * n_col):
        plt.subplot(n_row, n_col, i + 1)
        plt.imshow(images[:, i].reshape((h, w)), cmap=plt.cm.gray)
        plt.title("Eigenface %d" % (i + 1), size=12)
        plt.xticks(())
        plt.yticks(())
```

> This function is adapted from the LFW dataset example code in the scikit-learn documentation available at http://scikit-learn.org/stable/auto_examples/applications/face_recognition.html.

We will now use this function to plot the top 10 Eigenfaces:

```
plot_gallery(pcs, 50, 50)
```

This should display the following plot:

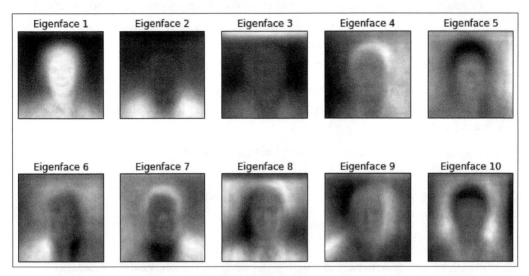

Top 10 Eigenfaces

Interpreting the Eigenfaces

Looking at the preceding images, we can see that the PCA model has effectively extracted recurring patterns of variation, which represent various features of the facial images. Each principal component can, as with clustering models, be interpreted. Again, like clustering, it is not always straightforward to interpret precisely what each principal component represents.

We can see from these images that there appear to be some images that pick up directional factors (for example, images 6 and 9), some hone in on hair patterns (such as images 4, 5, 7, and 10), while others seem to be somewhat more related to facial features such as eyes, nose, and mouth (such as images 1, 7, and 9).

Using a dimensionality reduction model

It is interesting to be able to visualize the outcome of a model in this way; however, the overall purpose of using dimensionality reduction is to create a more compact representation of the data that still captures the important features and variability in the raw dataset. To do this, we need to use a trained model to transform our raw data by projecting it into the new, lower-dimensional space represented by the principal components.

Projecting data using PCA on the LFW dataset

We will illustrate this concept by projecting each LFW image into a ten-dimensional vector. This is done through a matrix multiplication of the image matrix with the matrix of principal components. As the image matrix is a distributed MLlib `RowMatrix`, Spark takes care of distributing this computation for us through the `multiply` function:

```
val projected = matrix.multiply(pc)
println(projected.numRows, projected.numCols)
```

This will give you the following output:

`(1055,10)`

Observe that each image that was of dimension 2500 has been transformed into a vector of size 10. Let's take a look at the first few vectors:

```
println(projected.rows.take(5).mkString("\n"))
```

Here is the output:

```
[2648.9455749636277,1340.3713412351376,443.67380716760965,
-353.0021423043161,52.53102289832631,423.39861446944354,
413.8429065865399,-484.18122999722294,87.98862070273545,
-104.62720604921965]
[172.67735747311974,663.9154866829355,261.0575622447282,
-711.4857925259682,462.7663154755333,167.3082231097332,
-71.44832640530836,624.4911488194524,892.3209964031695,
-528.0056327351435]
[-1063.4562028554978,388.3510869550539,1508.2535609357597,
361.2485590837186,282.08588829583596,-554.3804376922453,
604.6680021092125,-224.16600191143075,-228.0771984153961,
-110.21539201855907]
[-4690.549692385103,241.83448841252638,-153.58903325799685,
-28.26215061165965,521.8908276360171,-442.0430200747375,
-490.1602309367725,-456.78026845649435,-78.79837478503592,
70.62925170688868]
[-2766.7960144161225,612.8408888724891,-405.76374113178616,
-468.56458995613974,863.1136863614743,-925.0935452709143,
69.24586949009642,-777.3348492244131,504.54033662376435,
257.0263568009851]
```

As the projected data is in the form of vectors, we can use the projection as input to another machine learning model. For example, we could use these projected inputs together with a set of input data generated from various images without faces to train a facial recognition model. Alternatively, we could train a multiclass classifier where each person is a class, thus creating a model that learns to identify the particular person that a face belongs to.

The relationship between PCA and SVD

We mentioned earlier that there is a close relationship between PCA and SVD. In fact, we can recover the same principal components and also apply the same projection into the space of principal components using SVD.

In our example, the right singular vectors derived from computing the SVD will be equivalent to the principal components we have calculated. We can see that this is the case by first computing the SVD on our image matrix and comparing the right singular vectors to the result of PCA. As was the case with PCA, SVD computation is provided as a function on a distributed `RowMatrix`:

```
val svd = matrix.computeSVD(10, computeU = true)
println(s"U dimension: (${svd.U.numRows}, ${svd.U.numCols})")
println(s"S dimension: (${svd.s.size}, )")
println(s"V dimension: (${svd.V.numRows}, ${svd.V.numCols})")
```

We can see that SVD returns a matrix `U` of dimension 1055 x 10, a vector `s` of the singular values of length `10`, and a matrix `V` of the right singular vectors of dimension 2500 x 10:

```
U dimension: (1055, 10)
S dimension: (10, )
V dimension: (2500, 10)
```

The matrix `V` is exactly equivalent to the result of PCA (ignoring the sign of the values and floating point tolerance). We can verify this with a utility function to compare the two by approximately comparing the data arrays of each matrix:

```
def approxEqual(array1: Array[Double], array2: Array[Double],
  tolerance: Double = 1e-6): Boolean = {
  // note we ignore sign of the principal component / singular
  vector elements
  val bools = array1.zip(array2).map { case (v1, v2) => if
  (math.abs(math.abs(v1) - math.abs(v2)) > 1e-6) false else true }
  bools.fold(true)(_ & _)
}
```

We will test the function on some test data:

```
println(approxEqual(Array(1.0, 2.0, 3.0), Array(1.0, 2.0, 3.0)))
```

This will give you the following output:

true

Let's try another test data:

```
println(approxEqual(Array(1.0, 2.0, 3.0), Array(3.0, 2.0, 1.0)))
```

This will give you the following output:

false

Finally, we can apply our equality function as follows:

```
println(approxEqual(svd.V.toArray, pc.toArray))
```

Here is the output:

true

The other relationship that holds is that the multiplication of the matrix U and vector S (or, strictly speaking, the diagonal matrix S) is equivalent to the PCA projection of our original image data into the space of the top 10 principal components.

We will now show that this is indeed the case. We will first use Breeze to multiply each vector in U by S, element-wise. We will then compare each vector in our PCA projected vectors with the equivalent vector in our SVD projection, and sum up the number of equal cases:

```
val breezeS = breeze.linalg.DenseVector(svd.s.toArray)
val projectedSVD = svd.U.rows.map { v =>
  val breezeV = breeze.linalg.DenseVector(v.toArray)
  val multV = breezeV :* breezeS
  Vectors.dense(multV.data)
}
projected.rows.zip(projectedSVD).map { case (v1, v2) =>
approxEqual(v1.toArray, v2.toArray) }.filter(b => true).count
```

This should display a result of 1055, as we would expect, confirming that each row of `projected` is equal to each row of `projectedSVD`.

 Note that the `:*` operator highlighted in the preceding code represents element-wise multiplication of the vectors.

Evaluating dimensionality reduction models

Both PCA and SVD are deterministic models. That is, given a certain input dataset, they will always produce the same result. This is in contrast to many of the models we have seen so far, which depend on some random element (most often for the initialization of model weight vectors and so on).

Both models are also guaranteed to return the top principal components or singular values, and hence, the only parameter is k. Like clustering models, increasing k always improves the model performance (for clustering, the relevant error function, while for PCA and SVD, the total amount of variability explained by the k components). Therefore, selecting a value for k is a trade-off between capturing as much structure of the data as possible while keeping the dimensionality of projected data low.

Evaluating k for SVD on the LFW dataset

We will examine the singular values obtained from computing the SVD on our image data. We can verify that the singular values are the same for each run and that they are returned in decreasing order, as follows:

```
val sValues = (1 to 5).map { i => matrix.computeSVD(i, computeU = false).s }
sValues.foreach(println)
```

This should show us output similar to the following:

```
[54091.00997110354]
[54091.00997110358,33757.702867982436]
[54091.00997110357,33757.70286798241,24541.193694775946]
[54091.00997110358,33757.70286798242,24541.19369477593,
23309.58418888302]
[54091.00997110358,33757.70286798242,24541.19369477593,
23309.584188882982,21803.09841158358]
```

As with evaluating values of k for clustering, in the case of SVD (and PCA), it is often useful to plot the singular values for a larger range of k and see where the point on the graph is where the amount of additional variance accounted for by each additional singular value starts to flatten out considerably.

We will do this by first computing the top 300 singular values:

```
val svd300 = matrix.computeSVD(300, computeU = false)
val sMatrix = new DenseMatrix(1, 300, svd300.s.toArray)
csvwrite(new File("/tmp/s.csv"), sMatrix)
```

We will write out the vector s of singular values to a temporary CSV file (as we did for our matrix of Eigenfaces previously) and then read it back in our IPython console, plotting the singular values for each k:

```
s = np.loadtxt("/tmp/s.csv", delimiter=",")
print(s.shape)
plot(s)
```

You should see an image displayed similar to the one shown here:

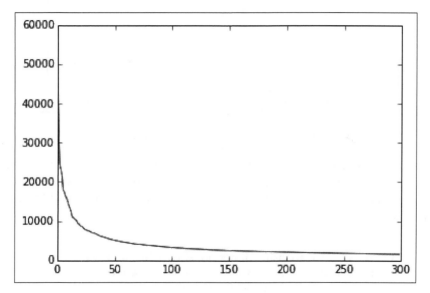

Top 300 singular values

A similar pattern is seen in the cumulative variation accounted for by the top 300 singular values (which we will plot on a log scale for the *y* axis):

```
plot(cumsum(s))
plt.yscale('log')
```

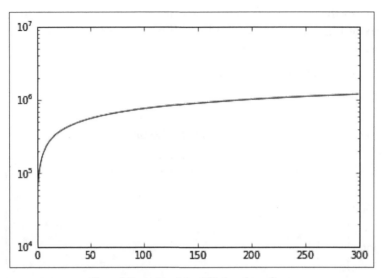

Cumulative sum of top 300 singular values

We can see that after a certain value range for *k* (around 100 in this case), the graph flattens considerably. This indicates that a number of singular values (or principal components) equivalent to this value of *k* probably explains enough of the variation of the original data.

> Of course, if we are using dimensionality reduction to help improve the performance of another model, we could use the same evaluation methods used for that model to help us choose a value for *k*.
>
> For example, we could use the AUC metric, together with cross-validation, to choose both the model parameters for a classification model as well as the value of *k* for our dimensionality reduction model. This does come at the expense of higher computation cost, however, as we would have to recompute the full model training and testing pipeline.

Summary

In this chapter, we explored two new unsupervised learning methods, PCA and SVD, for dimensionality reduction. We saw how to extract features for and train these models using facial image data. We visualized the results of the model in the form of Eigenfaces, saw how to apply the models to transform our original data into a reduced dimensionality representation, and investigated the close link between PCA and SVD.

In the next chapter, we will delve more deeply into techniques for text processing and analysis with Spark.

9
Advanced Text Processing with Spark

In *Chapter 3, Obtaining, Processing, and Preparing Data with Spark*, we covered various topics related to feature extraction and data processing, including the basics of extracting features from text data. In this chapter, we will introduce more advanced text processing techniques available in MLlib to work with large-scale text datasets.

In this chapter, we will:

- Work through detailed examples that illustrate data processing, feature extraction, and the modeling pipeline, as they relate to text data
- Evaluate the similarity between two documents based on the words in the documents
- Use the extracted text features as inputs for a classification model
- Cover a recent development in natural language processing to model words themselves as vectors and illustrate the use of Spark's **Word2Vec** model to evaluate the similarity between two words, based on their meaning

What's so special about text data?

Text data can be complex to work with for two main reasons. First, text and language have an inherent structure that is not easily captured using the raw words as is (for example, meaning, context, different types of words, sentence structure, and different languages, to highlight a few). Therefore, naïve feature extraction is usually relatively ineffective.

Second, the effective dimensionality of text data is extremely large and potentially limitless. Think about the number of words in the English language alone and add all kinds of special words, characters, slang, and so on to this. Then, throw in other languages and all the types of text one might find across the Internet. The dimension of text data can easily exceed tens or even hundreds of millions of words, even in relatively small datasets. For example, the Common Crawl dataset of billions of websites contains over 840 billion individual words.

To deal with these issues, we need ways of extracting more structured features and methods to handle the huge dimensionality of text data.

Extracting the right features from your data

The field of **natural language processing** (**NLP**) covers a wide range of techniques to work with text, from text processing and feature extraction through to modeling and machine learning. In this chapter, we will focus on two feature extraction techniques available within MLlib: the TF-IDF term weighting scheme and feature hashing.

Working through an example of TF-IDF, we will also explore the ways in which processing, tokenization, and filtering during feature extraction can help reduce the dimensionality of our input data as well as improve the information content and usefulness of the features we extract.

Term weighting schemes

In *Chapter 3, Obtaining, Processing, and Preparing Data with Spark*, we looked at vector representation where text features are mapped to a simple binary vector called the **bag-of-words** model. Another representation used commonly in practice is called **term frequency-inverse document frequency** (**TF-IDF**).

TF-IDF weights each term in a piece of text (referred to as a **document**) based on its frequency in the document (the **term frequency**). A global normalization, called the **inverse document frequency**, is then applied based on the frequency of this term among all documents (the set of documents in a dataset is commonly referred to as a **corpus**). The standard definition of TF-IDF is shown here:

$$\text{tf-idf}(t,d) = \text{tf}(t,d) \times \text{idf}(t)$$

Here, *tf(t,d)* is the frequency (number of occurrences) of term *t* in document *d* and *idf(t)* is the inverse document frequency of term *t* in the corpus; this is defined as follows:

$$idf(t) = \log(N / d)$$

Here, *N* is the total number of documents, and *d* is the number of documents in which the term *t* occurs.

The TF-IDF formulation means that terms occurring many times in a document receive a higher weighting in the vector representation relative to those that occur few times in the document. However, the IDF normalization has the effect of reducing the weight of terms that are very common across all documents. The end result is that truly rare or important terms should be assigned higher weighting, while more common terms (which are assumed to have less importance) should have less impact in terms of weighting.

> A good resource to learn more about the bag-of-words model (or **vector space model**) is the book *Introduction to Information Retrieval*, Christopher D. Manning, Prabhakar Raghavan and Hinrich Schütze, Cambridge University Press (available in HTML form at http://nlp.stanford.edu/IR-book/html/htmledition/irbook.html).
>
> It contains sections on text processing techniques, including tokenization, stop word removal, stemming, and the vector space model, as well as weighting schemes such as TF-IDF.
>
> An overview can also be found at http://en.wikipedia.org/wiki/Tf%E2%80%93idf.

Feature hashing

Feature hashing is a technique to deal with high-dimensional data and is often used with text and categorical datasets where the features can take on many unique values (often many millions of values). In the previous chapters, we often used the *1-of-K* encoding approach for categorical features, including text. While this approach is simple and effective, it can break down in the face of extremely high-dimensional data.

Building and using *1-of-K* feature encoding requires us to keep a mapping of each possible feature value to an index in a vector. Furthermore, the process of creating the mapping itself requires at least one additional pass through the dataset and can be tricky to do in parallel scenarios. Up until now, we have often used a simple approach of collecting the distinct feature values and zipping this collection with a set of indices to create a map of feature value to index. This mapping is then broadcast (either explicitly in our code or implicitly by Spark) to each worker.

However, when dealing with huge feature dimensions in the tens of millions or more that are common when working with text, this approach can be slow and can require significant memory and network resources, both on the Spark master (to collect the unique values) and workers (to broadcast the resulting mapping to each worker, which keeps it in memory to allow it to apply the feature encoding to its local piece of the input data).

Feature hashing works by assigning the vector index for a feature based on the value obtained by hashing this feature to a number (usually, an integer value) using a hash function. For example, let's say the hash value of a categorical feature for the geolocation of United States is 342. We will use the hashed value as the vector index, and the value at this index will be 1.0 to indicate the presence of the United States feature. The hash function used must be consistent (that is, for a given input, it returns the same output each time).

This encoding works the same way as mapping-based encoding, except that we choose a size for our feature vector upfront. As the most common hash functions return values in the entire range of integers, we will use a *modulo* operation to restrict the index values to the size of our vector, which is typically much smaller (a few tens of thousands to a few million, depending on our requirements).

Feature hashing has the advantage that we do not need to build a mapping and keep it in memory. It is also easy to implement, very fast, and can be done online and in real time, thus not requiring a pass through our dataset first. Finally, because we selected a feature vector dimension that is significantly smaller than the raw dimensionality of our dataset, we bound the memory usage of our model both in training and production; hence, memory usage does not scale with the size and dimensionality of our data.

However, there are two important drawbacks, which are as follows:

- As we don't create a mapping of features to index values, we also cannot do the reverse mapping of feature index to value. This makes it harder to, for example, determine which features are most informative in our models.
- As we are restricting the size of our feature vectors, we might experience **hash collisions**. This happens when two different features are hashed into the same index in our feature vector. Surprisingly, this doesn't seem to have a severe impact on model performance as long as we choose a reasonable feature vector dimension relative to the dimension of the input data.

> Further information on hashing can be found at http://en.wikipedia.org/wiki/Hash_function.
>
> A key paper that introduced the use of hashing for feature extraction and machine learning is:
>
> *Kilian Weinberger, Anirban Dasgupta, John Langford, Alex Smola,* and *Josh Attenberg. Feature Hashing for Large Scale Multitask Learning. Proc. ICML 2009,* which is available at http://alex.smola.org/papers/2009/Weinbergeretal09.pdf.

Extracting the TF-IDF features from the 20 Newsgroups dataset

To illustrate the concepts in this chapter, we will use a well-known text dataset called **20 Newsgroups**; this dataset is commonly used for text-classification tasks. This is a collection of newsgroup messages posted across 20 different topics. There are various forms of data available. For our purposes, we will use the bydate version of the dataset, which is available at http://qwone.com/~jason/20Newsgroups.

This dataset splits up the available data into training and test sets that comprise 60 percent and 40 percent of the original data, respectively. Here, the messages in the test set occur after those in the training set. This dataset also excludes some of the message headers that identify the actual newsgroup; hence, it is an appropriate dataset to test the real-world performance of classification models.

> Further information on the original dataset can be found in the *UCI Machine Learning Repository* page at http://kdd.ics.uci.edu/databases/20newsgroups/20newsgroups.data.html.

To get started, download the data and unzip the file using the following command:

```
>tar xfvz 20news-bydate.tar.gz
```

This will create two folders: one called 20news-bydate-train and another one called 20news-bydate-test. Let's take a look at the directory structure under the training dataset folder:

```
>cd 20news-bydate-train/
>ls
```

Advanced Text Processing with Spark

You will see that it contains a number of subfolders, one for each newsgroup:

```
alt.atheism                comp.windows.x         rec.sport.hockey
soc.religion.christian
comp.graphics              misc.forsale           sci.crypt
talk.politics.guns
comp.os.ms-windows.misc    rec.autos              sci.electronics
talk.politics.mideast
comp.sys.ibm.pc.hardware   rec.motorcycles        sci.med
talk.politics.misc
comp.sys.mac.hardware      rec.sport.baseball     sci.space
talk.religion.misc
```

There are a number of files under each newsgroup folder; each file contains an individual message posting:

```
> ls rec.sport.hockey
52550 52580 52610 52640 53468 53550 53580 53610 53640 53670 53700
53731 53761 53791
```

...

We can take a look at a part of one of these messages to see the format:

```
> head -20 rec.sport.hockey/52550
From: dchhabra@stpl.ists.ca (Deepak Chhabra)
Subject: Superstars and attendance (was Teemu Selanne, was +/-
leaders)
Nntp-Posting-Host: stpl.ists.ca
Organization: Solar Terresterial Physics Laboratory, ISTS
Distribution: na
Lines: 115


Dean J. Falcione (posting from jrmst+8@pitt.edu) writes:
[I wrote:]

>>When the Pens got Mario, granted there was big publicity, etc, etc,
>>and interest was immediately generated.  Gretzky did the same thing
for LA.
```

```
>>However, imnsho, neither team would have seen a marked improvement
in
>>attendance if the team record did not improve.  In the year before
Lemieux
>>came, Pittsburgh finished with 38 points.  Following his arrival,
the Pens
>>finished with 53, 76, 72, 81, 87, 72, 88, and 87 points, with a
couple of
                                    ^^
>>Stanley Cups thrown in.
...
```

As we can see, each message contains some header fields that contain the sender, subject, and other metadata, followed by the raw content of the message.

Exploring the 20 Newsgroups data

Now, we will start up our Spark Scala console, ensuring that we make enough memory available:

```
>./SPARK_HOME/bin/spark-shell --driver-memory 4g
```

Looking at the directory structure, you might recognize that once again, we have data contained in individual text files (one text file per message). Therefore, we will again use Spark's `wholeTextFiles` method to read the content of each file into a record in our RDD.

In the code that follows, PATH refers to the directory in which you extracted the `20news-bydate` ZIP file:

```
val path = "/PATH/20news-bydate-train/*"
val rdd = sc.wholeTextFiles(path)
val text = rdd.map { case (file, text) => text }
println(text.count)
```

The first time you run this command, it might take quite a bit of time, as Spark needs to scan the directory structure. You will also see quite a lot of console output, as Spark logs all the file paths that are being processed. During the processing, you will see the following line displayed, indicating the total number of files that Spark has detected:

```
...
14/10/12 14:27:54 INFO FileInputFormat: Total input paths to process
: 11314
...
```

After the command has finished running, you will see the total record count, which should be the same as the preceding **Total input paths to process** screen output:

```
11314
```

Next, we will take a look at the newsgroup topics available:

```
val newsgroups = rdd.map { case (file, text) =>
file.split("/").takeRight(2).head }
val countByGroup = newsgroups.map(n => (n, 1)).reduceByKey
(_ + _).collect.sortBy(-_._2).mkString("\n")
println(countByGroup)
```

This will display the following result:

```
(rec.sport.hockey,600)
(soc.religion.christian,599)
(rec.motorcycles,598)
(rec.sport.baseball,597)
(sci.crypt,595)
(rec.autos,594)
(sci.med,594)
(comp.windows.x,593)
(sci.space,593)
(sci.electronics,591)
(comp.os.ms-windows.misc,591)
(comp.sys.ibm.pc.hardware,590)
(misc.forsale,585)
(comp.graphics,584)
(comp.sys.mac.hardware,578)
(talk.politics.mideast,564)
(talk.politics.guns,546)
(alt.atheism,480)
(talk.politics.misc,465)
(talk.religion.misc,377)
```

We can see that the number of messages is roughly even between the topics.

Applying basic tokenization

The first step in our text processing pipeline is to split up the raw text content in each document into a collection of terms (also referred to as **tokens**). This is known as **tokenization**. We will start by applying a simple **whitespace** tokenization, together with converting each token to lowercase for each document:

```
val text = rdd.map { case (file, text) => text }
val whiteSpaceSplit = text.flatMap(t => t.split(" ").map(_.
toLowerCase))
println(whiteSpaceSplit.distinct.count)
```

> In the preceding code, we used the `flatMap` function instead of `map`, as for now, we want to inspect all the tokens together for exploratory analysis. Later in this chapter, we will apply our tokenization scheme on a per-document basis, so we will use the `map` function.

After running this code snippet, you will see the total number of unique tokens after applying our tokenization:

`402978`

As you can see, for even a relatively small set of text, the number of raw tokens (and, therefore, the dimensionality of our feature vectors) can be very high.

Let's take a look at a randomly selected document:

```
println(whiteSpaceSplit.sample
(true, 0.3, 42).take(100).mkString(","))
```

> Note that we set the third parameter to the `sample` function, which is the random seed. We set this function to `42` so that we get the same results from each call to `sample` so that your results match those in this chapter.

This will display the following result:

```
atheist,resources
summary:,addresses,,to,atheism
keywords:,music,,thu,,11:57:19,11:57:19,gmt
distribution:,cambridge.,290
```

Advanced Text Processing with Spark

```
archive-name:,atheism/resources
alt-atheism-archive-name:,december,,,,,,,,,,,,,,,,,,,,addresses,address
es,,,,,,,
religion,to:,to:,,p.o.,53701.
telephone:,sell,the,,fish,on,their,cars,,with,and,written
inside.,3d,plastic,plastic,,evolution,evolution,7119,,,,,san,san,san,
mailing,net,who,to,atheist,press


aap,various,bible,,and,on.,,,one,book,is:


"the,w.p.,american,pp.,,1986.,bible,contains,ball,,based,based,james,
of
```

Improving our tokenization

The preceding simple approach results in a lot of tokens and does not filter out many nonword characters (such as punctuation). Most tokenization schemes will remove these characters. We can do this by splitting each raw document on **nonword characters** using a regular expression pattern:

```
val nonWordSplit = text.flatMap(t =>
t.split("""\W+""").map(_.toLowerCase))
println(nonWordSplit.distinct.count)
```

This reduces the number of unique tokens significantly:

```
130126
```

If we inspect the first few tokens, we will see that we have eliminated most of the less useful characters in the text:

```
println(nonWordSplit.distinct.sample
(true, 0.3, 42).take(100).mkString(","))
```

You will see the following result displayed:

```
bone,k29p,w1w3s1,odwyer,dnj33n,bruns,_congressional,mmejv5,mmejv5,art
ur,125215,entitlements,beleive,1pqd9hinnbmi,
jxicaijp,b0vp,underscored,believiing,qsins,1472,urtfi,nauseam,tohc4,k
ielbasa,ao,wargame,seetex,museum,typeset,pgva4,
dcbq,ja_jp,ww4ewa4g,animating,animating,10011100b,10011100b,413,wp3d,
wp3d,cannibal,searflame,ets,1qjfnv,6jx,6jx,
detergent,yan,aanp,unaskable,9mf,bowdoin,chov,16mb,createwindow,kjznk
h,df,classifieds,hour,cfsmo,santiago,santiago,
```

1r1d62,almanac_,almanac_,chq,nowadays,formac,formac,bacteriophage,bar
king,barking,barking,ipmgocj7b,monger,projector,

hama,65e90h8y,homewriter,c15,1496,zysec,homerific,00ecgillespie,00ecg
illespie,mqh0,suspects,steve_mullins,io21087,

funded,liberated,canonical,throng,0hnz,exxon,xtappcontext,mcdcup,mcdc
up,5seg,biscuits

While our nonword pattern to split text works fairly well, we are still left with numbers and tokens that contain numeric characters. In some cases, numbers can be an important part of a corpus. For our purposes, the next step in our pipeline will be to filter out numbers and tokens that are words mixed with numbers.

We can do this by applying another regular expression pattern and using this to filter out tokens that *do not match* the pattern:

```
val regex = """[^0-9]*""".r
val filterNumbers = nonWordSplit.filter(token =>
regex.pattern.matcher(token).matches)
println(filterNumbers.distinct.count)
```

This further reduces the size of the token set:

84912

Let's take a look at another random sample of the filtered tokens:

```
println(filterNumbers.distinct.sample
(true, 0.3, 42).take(100).mkString(","))
```

You will see output like the following one:

reunion,wuair,schwabam,eer,silikian,fuller,sloppiness,crying,crying,
beckmans,leymarie,fowl,husky,rlhzrlhz,ignore,

loyalists,goofed,arius,isgal,dfuller,neurologists,robin,jxicaijp,
majorly,nondiscriminatory,akl,sively,adultery,

urtfi,kielbasa,ao,instantaneous,subscriptions,collins,collins,za_,za_
,jmckinney,nonmeasurable,nonmeasurable,

seetex,kjvar,dcbq,randall_clark,theoreticians,theoreticians,
congresswoman,sparcstaton,diccon,nonnemacher,

arresed,ets,sganet,internship,bombay,keysym,newsserver,connecters,
igpp,aichi,impute,impute,raffle,nixdorf,

nixdorf,amazement,butterfield,geosync,geosync,scoliosis,eng,eng,eng,
kjznkh,explorers,antisemites,bombardments,

```
abba,caramate,tully,mishandles,wgtn,springer,nkm,nkm,alchoholic,chq,
shutdown,bruncati,nowadays,mtearle,eastre,
```
```
discernible,bacteriophage,paradijs,systematically,rluap,rluap,blown,
moderates
```

We can see that we have removed all the numeric characters. This still leaves us with a few strange *words,* but we will not worry about these too much here.

Removing stop words

Stop words refer to common words that occur many times across almost all documents in a corpus (and across most corpuses). Examples of typical English stop words include and, but, the, of, and so on. It is a standard practice in text feature extraction to exclude stop words from the extracted tokens.

When using TF-IDF weighting, the weighting scheme actually takes care of this for us. As stop words have a very low IDF score, they will tend to have very low TF-IDF weightings and thus less importance. In some cases, for information retrieval and search tasks, it might be desirable to include stop words. However, it can still be beneficial to exclude stop words during feature extraction, as it reduces the dimensionality of the final feature vectors as well as the size of the training data.

We can take a look at some of the tokens in our corpus that have the highest occurrence across all documents to get an idea about some other stop words to exclude:

```
val tokenCounts = filterNumbers.map(t => (t, 1)).reduceByKey
(_ + _)
val oreringDesc = Ordering.by[(String, Int), Int](_._2)
println(tokenCounts.top(20)(oreringDesc).mkString("\n"))
```

In the preceding code, we took the tokens after filtering out numeric characters and generated a count of the occurrence of each token across the corpus. We can now use Spark's `top` function to retrieve the top 20 tokens by count. Notice that we need to provide the `top` function with an ordering that tells Spark how to order the elements of our RDD. In this case, we want to order by the count, so we will specify the second element of our key-value pair.

Running the preceding code snippet will result in the following top tokens:

```
(the,146532)
(to,75064)
(of,69034)
(a,64195)
(ax,62406)
```

(and,57957)

(i,53036)

(in,49402)

(is,43480)

(that,39264)

(it,33638)

(for,28600)

(you,26682)

(from,22670)

(s,22337)

(edu,21321)

(on,20493)

(this,20121)

(be,19285)

(t,18728)

As we might expect, there are a lot of common words in this list that we could potentially label as stop words. Let's create a set of stop words with some of these as well as other common words. We will then look at the tokens after filtering out these stop words:

```
val stopwords = Set(
  "the","a","an","of","or","in","for","by","on","but", "is", "not",
"with", "as", "was", "if",
  "they", "are", "this", "and", "it", "have", "from", "at", "my",
"be", "that", "to"
)
val tokenCountsFilteredStopwords = tokenCounts.filter { case
(k, v) => !stopwords.contains(k) }
println(tokenCountsFilteredStopwords.top(20)(oreringDesc).mkString
("\n"))
```

You will see the following output:

(ax,62406)

(i,53036)

(you,26682)

(s,22337)

(edu,21321)

(t,18728)

```
(m,12756)
(subject,12264)
(com,12133)
(lines,11835)
(can,11355)
(organization,11233)
(re,10534)
(what,9861)
(there,9689)
(x,9332)
(all,9310)
(will,9279)
(we,9227)
(one,9008)
```

You might notice that there are still quite a few common words in this top list. In practice, we might have a much larger set of stop words. However, we will keep a few (partly to illustrate the impact of common words when using TF-IDF weighting a little later).

One other filtering step that we will use is removing any tokens that are only one character in length. The reasoning behind this is similar to removing stop words—these single-character tokens are unlikely to be informative in our text model and can further reduce the feature dimension and model size. We will do this with another filtering step:

```
val tokenCountsFilteredSize = tokenCountsFilteredStopwords.filter
{ case (k, v) => k.size >= 2 }
println(tokenCountsFilteredSize.top(20)(oreringDesc).mkString
("\n"))
```

Again, we will examine the tokens remaining after this filtering step:

```
(ax,62406)
(you,26682)
(edu,21321)
(subject,12264)
(com,12133)
(lines,11835)
(can,11355)
```

```
(organization,11233)
(re,10534)
(what,9861)
(there,9689)
(all,9310)
(will,9279)
(we,9227)
(one,9008)
(would,8905)
(do,8674)
(he,8441)
(about,8336)
(writes,7844)
```

Apart from some of the common words that we have not excluded, we see that a few potentially more informative words are starting to appear.

Excluding terms based on frequency

It is also a common practice to exclude terms during tokenization when their overall occurrence in the corpus is very low. For example, let's examine the least occurring terms in the corpus (notice the different ordering we use here to return the results sorted in ascending order):

```
val oreringAsc = Ordering.by[(String, Int), Int](-_._2)
println(tokenCountsFilteredSize.top(20)(oreringAsc).mkString
("\n"))
```

You will get the following results:

```
(lennips,1)
(bluffing,1)
(preload,1)
(altina,1)
(dan_jacobson,1)
(vno,1)
(actu,1)
(donnalyn,1)
(ydag,1)
```

```
(mirosoft,1)
(xiconfiywindow,1)
(harger,1)
(feh,1)
(bankruptcies,1)
(uncompression,1)
(d_nibby,1)
(bunuel,1)
(odf,1)
(swith,1)
(lantastic,1)
```

As we can see, there are many terms that only occur once in the entire corpus. Since typically we want to use our extracted features for other tasks such as document similarity or machine learning models, tokens that only occur once are not useful to learn from, as we will not have enough training data relative to these tokens. We can apply another filter to exclude these rare tokens:

```
val rareTokens = tokenCounts.filter{ case (k, v) => v < 2 }.map {
case (k, v) => k }.collect.toSet
val tokenCountsFilteredAll = tokenCountsFilteredSize.filter { case
(k, v) => !rareTokens.contains(k) }
println(tokenCountsFilteredAll.top(20)(oreringAsc).mkString("\n"))
```

We can see that we are left with tokens that occur at least twice in the corpus:

```
(sina,2)
(akachhy,2)
(mvd,2)
(hizbolah,2)
(wendel_clark,2)
(sarkis,2)
(purposeful,2)
(feagans,2)
(wout,2)
(uneven,2)
(senna,2)
(multimeters,2)
(bushy,2)
```

```
(subdivided,2)
(coretest,2)
(oww,2)
(historicity,2)
(mmg,2)
(margitan,2)
(defiance,2)
```

Now, let's count the number of unique tokens:

```
println(tokenCountsFilteredAll.count)
```

You will see the following output:

```
51801
```

As we can see, by applying all the filtering steps in our tokenization pipeline, we have reduced the feature dimension from 402,978 to 51,801.

We can now combine all our filtering logic into one function, which we can apply to each document in our RDD:

```
def tokenize(line: String): Seq[String] = {
  line.split("""\W+""")
    .map(_.toLowerCase)
    .filter(token => regex.pattern.matcher(token).matches)
    .filterNot(token => stopwords.contains(token))
    .filterNot(token => rareTokens.contains(token))
    .filter(token => token.size >= 2)
    .toSeq
}
```

We can check whether this function gives us the same result with the following code snippet:

```
println(text.flatMap(doc => tokenize(doc)).distinct.count)
```

This will output `51801`, giving us the same unique token count as our step-by-step pipeline.

We can tokenize each document in our RDD as follows:

```
val tokens = text.map(doc => tokenize(doc))
println(tokens.first.take(20))
```

You will see output similar to the following, showing the first part of the tokenized version of our first document:

```
WrappedArray(mathew, mathew, mantis, co, uk, subject, alt, atheism,
faq, atheist, resources, summary, books, addresses, music, anything,
related, atheism, keywords, faq)
```

A note about stemming

A common step in text processing and tokenization is **stemming**. This is the conversion of whole words to a **base form** (called a **word stem**). For example, plurals might be converted to singular (*dogs* becomes *dog*), and forms such as *walking* and *walker* might become *walk*. Stemming can become quite complex and is typically handled with specialized NLP or search engine software (such as NLTK, OpenNLP, and Lucene, for example). We will ignore stemming for the purpose of our example here.

 A full treatment of stemming is beyond the scope of this book. You can find more details at http://en.wikipedia.org/wiki/Stemming.

Training a TF-IDF model

We will now use MLlib to transform each document, in the form of processed tokens, into a vector representation. The first step will be to use the `HashingTF` implementation, which makes use of feature hashing to map each token in the input text to an index in the vector of term frequencies. Then, we will compute the global IDF and use it to transform the term frequency vectors into TF-IDF vectors.

For each token, the index will thus be the hash of the token (mapped in turn onto the dimension of the feature vector). The value for each token will be the TF-IDF weighting for that token (that is, the term frequency multiplied by the inverse document frequency).

First, we will import the classes we need and create our `HashingTF` instance, passing in a `dim` dimension parameter. While the default feature dimension is 2^{20} (or around 1 million), we will choose 2^{18} (or around 260,000), since with about 50,000 tokens, we should not experience a significant number of hash collisions, and a smaller dimension will be more memory and processing friendly for illustrative purposes:

```
import org.apache.spark.mllib.linalg.{ SparseVector => SV }
import org.apache.spark.mllib.feature.HashingTF
import org.apache.spark.mllib.feature.IDF
```

```
val dim = math.pow(2, 18).toInt
val hashingTF = new HashingTF(dim)
val tf = hashingTF.transform(tokens)
tf.cache
```

Note that we imported MLlib's `SparseVector` using an alias of `SV`. This is because later, we will use Breeze's `linalg` module, which itself also imports `SparseVector`. This way, we will avoid namespace collisions.

The `transform` function of `HashingTF` maps each input document (that is, a sequence of tokens) to an MLlib `Vector`. We will also call `cache` to pin the data in memory to speed up subsequent operations.

Let's inspect the first element of our transformed dataset:

Note that `HashingTF.transform` returns an `RDD[Vector]`, so we will cast the result returned to an instance of an MLlib `SparseVector`.

The `transform` method can also work on an individual document by taking an `Iterable` argument (for example, a document as a `Seq[String]`). This returns a single vector.

```
val v = tf.first.asInstanceOf[SV]
println(v.size)
println(v.values.size)
println(v.values.take(10).toSeq)
println(v.indices.take(10).toSeq)
```

You will see the following output displayed:

262144

706

WrappedArray(1.0, 1.0, 1.0, 1.0, 2.0, 1.0, 1.0, 2.0, 1.0, 1.0)

WrappedArray(313, 713, 871, 1202, 1203, 1209, 1795, 1862, 3115, 3166)

We can see that the dimension of each sparse vector of term frequencies is 262,144 (or 2^{18} as we specified). However, the number on non-zero entries in the vector is only 706. The last two lines of the output show the frequency counts and vector indexes for the first few entries in the vector.

We will now compute the inverse document frequency for each term in the corpus by creating a new `IDF` instance and calling `fit` with our RDD of term frequency vectors as the input. We will then transform our term frequency vectors to TF-IDF vectors through the `transform` function of `IDF`:

```
val idf = new IDF().fit(tf)
val tfidf = idf.transform(tf)
val v2 = tfidf.first.asInstanceOf[SV]
println(v2.values.size)
println(v2.values.take(10).toSeq)
println(v2.indices.take(10).toSeq)
```

When you examine the first element in the RDD of TF-IDF transformed vectors, you will see output similar to the one shown here:

```
706
WrappedArray(2.3869085659322193, 4.670445463955571,
6.561295835827856, 4.597686109673142, ...
WrappedArray(313, 713, 871, 1202, 1203, 1209, 1795, 1862, 3115, 3166)
```

We can see that the number of non-zero entries hasn't changed (at 706), nor have the vector indices for the terms. What has changed are the values for each term. Earlier, these represented the frequency of each term in the document, but now, the new values represent the frequencies weighted by the `IDF`.

Analyzing the TF-IDF weightings

Next, let's investigate the TF-IDF weighting for a few terms to illustrate the impact of the commonality or rarity of a term.

First, we can compute the minimum and maximum TF-IDF weights across the entire corpus:

```
val minMaxVals = tfidf.map { v =>
  val sv = v.asInstanceOf[SV]
  (sv.values.min, sv.values.max)
}
val globalMinMax = minMaxVals.reduce { case ((min1, max1),
(min2, max2)) =>
  (math.min(min1, min2), math.max(max1, max2))
}
println(globalMinMax)
```

As we can see, the minimum TF-IDF is zero, while the maximum is significantly larger:

```
(0.0,66155.39470409753)
```

We will now explore the TF-IDF weight attached to various terms. In the previous section on stop words, we filtered out many common terms that occur frequently. Recall that we did not remove all such potential stop words. Instead, we kept a few in the corpus so that we could illustrate the impact of applying the TF-IDF weighting scheme on these terms.

TF-IDF weighting will tend to assign a lower weighting to common terms. To see this, we can compute the TF-IDF representation for a few of the terms that appear in the list of top occurrences that we previously computed, such as you, do, and we:

```
val common = sc.parallelize(Seq(Seq("you", "do", "we")))
val tfCommon = hashingTF.transform(common)
val tfidfCommon = idf.transform(tfCommon)
val commonVector = tfidfCommon.first.asInstanceOf[SV]
println(commonVector.values.toSeq)
```

If we form a TF-IDF vector representation of this document, we would see the following values assigned to each term. Note that because of feature hashing, we are not sure exactly which term represents what. However, the values illustrate that the weighting applied to these terms is relatively low:

```
WrappedArray(0.9965359935704624, 1.3348773448236835,
0.5457486182039175)
```

Now, let's apply the same transformation to a few less common terms that we might intuitively associate with being more linked to specific topics or concepts:

```
val uncommon = sc.parallelize(Seq(Seq("telescope", "legislation",
"investment")))
val tfUncommon = hashingTF.transform(uncommon)
val tfidfUncommon = idf.transform(tfUncommon)
val uncommonVector = tfidfUncommon.first.asInstanceOf[SV]
println(uncommonVector.values.toSeq)
```

We can see from the following results that the TF-IDF weightings are indeed significantly higher than for the more common terms:

```
WrappedArray(5.3265513728351666, 5.308532867332488,
5.483736956357579)
```

Using a TF-IDF model

While we often refer to training a TF-IDF model, it is actually a feature extraction process or transformation rather than a machine learning model. TF-IDF weighting is often used as a preprocessing step for other models, such as dimensionality reduction, classification, or regression.

To illustrate the potential uses of TF-IDF weighting, we will explore two examples. The first is using the TF-IDF vectors to compute document similarity, while the second involves training a multilabel classification model with the TF-IDF vectors as input features.

Document similarity with the 20 Newsgroups dataset and TF-IDF features

You might recall from *Chapter 4, Building a Recommendation Engine with Spark*, that the similarity between two vectors can be computed using a distance metric. The closer two vectors are (that is, the lower the distance metric), the more similar they are. One such metric that we used to compute similarity between movies is cosine similarity.

Just like we did for movies, we can also compute the similarity between two documents. Using TF-IDF, we have transformed each document into a vector representation. Hence, we can use the same techniques as we used for movie vectors to compare two documents.

Intuitively, we might expect two documents to be more similar to each other if they share many terms. Conversely, we might expect two documents to be less similar if they each contain many terms that are different from each other. As we compute cosine similarity by computing a dot product of the two vectors and each vector is made up of the terms in each document, we can see that documents with a high overlap of terms will tend to have a higher cosine similarity.

Now, we can see TF-IDF at work. We might reasonably expect that even very different documents might contain many overlapping terms that are relatively common (for example, our stop words). However, due to a low TF-IDF weighting, these terms will not have a significant impact on the dot product and, therefore, will not have much impact on the similarity computed.

For example, we might expect two randomly chosen messages from the `hockey` newsgroup to be relatively similar to each other. Let's see if this is the case:

```
val hockeyText = rdd.filter { case (file, text) =>
file.contains("hockey") }
```

```
val hockeyTF = hockeyText.mapValues(doc =>
hashingTF.transform(tokenize(doc)))
val hockeyTfIdf = idf.transform(hockeyTF.map(_._2))
```

In the preceding code, we first filtered our raw input RDD to keep only the messages within the hockey topic. We then applied our tokenization and term frequency transformation functions. Note that the `transform` method used is the version that works on a single document (in the form of a `Seq[String]`) rather than the version that works on an RDD of documents.

Finally, we applied the `IDF` transform (note that we use the same IDF that we have already computed on the whole corpus).

Once we have our `hockey` document vectors, we can select two of these vectors at random and compute the cosine similarity between them (as we did earlier, we will use Breeze for the linear algebra functionality, in particular converting our MLlib vectors to Breeze `SparseVector` instances first):

```
import breeze.linalg._
val hockey1 = hockeyTfIdf.sample
(true, 0.1, 42).first.asInstanceOf[SV]
val breeze1 = new SparseVector(hockey1.indices, hockey1.values,
hockey1.size)
val hockey2 = hockeyTfIdf.sample
(true, 0.1, 43).first.asInstanceOf[SV]
val breeze2 = new SparseVector(hockey2.indices, hockey2.values,
hockey2.size)
val cosineSim = breeze1.dot(breeze2) / (norm(breeze1) *
norm(breeze2))
println(cosineSim)
```

We can see that the cosine similarity between the documents is around 0.06:

`0.060250114361164626`

While this might seem quite low, recall that the effective dimensionality of our features is high due to the large number of unique terms that is typical when dealing with text data. Hence, we can expect that any two documents might have a relatively low overlap of terms even if they are about the same topic, and therefore would have a lower absolute similarity score.

By contrast, we can compare this similarity score to the one computed between one of our `hockey` documents and another document chosen randomly from the `comp.graphics` newsgroup, using the same methodology:

```
val graphicsText = rdd.filter { case (file, text) =>
file.contains("comp.graphics") }
val graphicsTF = graphicsText.mapValues(doc =>
hashingTF.transform(tokenize(doc)))
val graphicsTfIdf = idf.transform(graphicsTF.map(_._2))
val graphics = graphicsTfIdf.sample
(true, 0.1, 42).first.asInstanceOf[SV]
val breezeGraphics = new SparseVector(graphics.indices,
graphics.values, graphics.size)
val cosineSim2 = breeze1.dot(breezeGraphics) / (norm(breeze1) *
norm(breezeGraphics))
println(cosineSim2)
```

The cosine similarity is significantly lower at 0.0047:

`0.004664850323792852`

Finally, it is likely that a document from another sports-related topic might be more similar to our `hockey` document than one from a computer-related topic. However, we would probably expect a `baseball` document to not be as similar as our `hockey` document. Let's see whether this is the case by computing the similarity between a random message from the `baseball` newsgroup and our `hockey` document:

```
val baseballText = rdd.filter { case (file, text) =>
file.contains("baseball") }
val baseballTF = baseballText.mapValues(doc =>
hashingTF.transform(tokenize(doc)))
val baseballTfIdf = idf.transform(baseballTF.map(_._2))
val baseball = baseballTfIdf.sample
(true, 0.1, 42).first.asInstanceOf[SV]
val breezeBaseball = new SparseVector(baseball.indices,
baseball.values, baseball.size)
val cosineSim3 = breeze1.dot(breezeBaseball) / (norm(breeze1) *
norm(breezeBaseball))
println(cosineSim3)
```

Indeed, as we expected, we found that the `baseball` and `hockey` documents have a cosine similarity of 0.05, which is significantly higher than the `comp.graphics` document, but also somewhat lower than the other `hockey` document:

`0.05047395039466008`

Training a text classifier on the 20 Newsgroups dataset using TF-IDF

When using TF-IDF vectors, we expected that the cosine similarity measure would capture the similarity between documents, based on the overlap of terms between them. In a similar way, we would expect that a machine learning model, such as a classifier, would be able to learn weightings for individual terms; this would allow it to distinguish between documents from different classes. That is, it should be possible to learn a mapping between the presence (and weighting) of certain terms and a specific topic.

In the 20 Newsgroups example, each newsgroup topic is a class, and we can train a classifier using our TF-IDF transformed vectors as input.

Since we are dealing with a multiclass classification problem, we will use the naïve Bayes model in MLlib, which supports multiple classes. As the first step, we will import the Spark classes that we will be using:

```
import org.apache.spark.mllib.regression.LabeledPoint
import org.apache.spark.mllib.classification.NaiveBayes
import org.apache.spark.mllib.evaluation.MulticlassMetrics
```

Next, we will need to extract the 20 topics and convert them to class mappings. We can do this in exactly the same way as we might for 1-of-K feature encoding, by assigning a numeric index to each class:

```
val newsgroupsMap =
newsgroups.distinct.collect().zipWithIndex.toMap
val zipped = newsgroups.zip(tfidf)
val train = zipped.map { case (topic, vector) =>
LabeledPoint(newsgroupsMap(topic), vector) }
train.cache
```

In the preceding code snippet, we took the `newsgroups` RDD, where each element is the topic, and used the `zip` function to combine it with each element in our `tfidf` RDD of TF-IDF vectors. We then mapped over each key-value element in our new `zipped` RDD and created a `LabeledPoint` instance, where `label` is the class index and `features` is the TF-IDF vector.

> Note that the `zip` operator assumes that each RDD has the same number of partitions as well as the same number of elements in each partition. It will fail if this is not the case. We can make this assumption here because we have effectively created both our `tfidf` RDD and `newsgroups` RDD from a series of `map` transformations on the same original RDD that preserved the partitioning structure.

Now that we have an input RDD in the correct form, we can simply pass it to the naïve Bayes `train` function:

```
val model = NaiveBayes.train(train, lambda = 0.1)
```

Let's evaluate the performance of the model on the test dataset. We will load the raw test data from the `20news-bydate-test` directory, again using `wholeTextFiles` to read each message into an RDD element. We will then extract the class labels from the file paths in the same way as we did for the `newsgroups` RDD:

```
val testPath = "/PATH/20news-bydate-test/*"
val testRDD = sc.wholeTextFiles(testPath)
val testLabels = testRDD.map { case (file, text) =>
  val topic = file.split("/").takeRight(2).head
  newsgroupsMap(topic)
}
```

Transforming the text in the test dataset follows the same procedure as for the training data—we will apply our `tokenize` function followed by the term frequency transformation, and we will again use the same IDF computed from the training data to transform the TF vectors into TF-IDF vectors. Finally, we will zip the test class labels with the TF-IDF vectors and create our test `RDD[LabeledPoint]`:

```
val testTf = testRDD.map { case (file, text) =>
hashingTF.transform(tokenize(text)) }
val testTfIdf = idf.transform(testTf)
val zippedTest = testLabels.zip(testTfIdf)
val test = zippedTest.map { case (topic, vector) =>
LabeledPoint(topic, vector) }
```

> Note that it is important that we use the training set IDF to transform the test data, as this creates a more realistic estimation of model performance on new data, which might potentially contain terms that the model has not yet been trained on. It would be "cheating" to recompute the IDF vector based on the test dataset and, more importantly, would potentially lead to incorrect estimates of optimal model parameters selected through cross-validation.

Now, we're ready to compute the predictions and true class labels for our model. We will use this RDD to compute accuracy and the multiclass weighted F-measure for our model:

```
val predictionAndLabel = test.map(p => (model.predict(p.features),
p.label))
val accuracy = 1.0 * predictionAndLabel.filter
(x => x._1 == x._2).count() / test.count()
val metrics = new MulticlassMetrics(predictionAndLabel)
println(accuracy)
println(metrics.weightedFMeasure)
```

> The weighted F-measure is an overall measure of precision and recall performance (where, like area under an ROC curve, values closer to 1.0 indicate better performance), which is then combined through a weighted averaged across the classes.

We can see that our simple multiclass naïve Bayes model has achieved close to 80 percent for both accuracy and F-measure:

0.7915560276155071

0.7810675969031116

Evaluating the impact of text processing

Text processing and TF-IDF weighting are examples of feature extraction techniques designed to both reduce the dimensionality of and extract some structure from raw text data. We can see the impact of applying these processing techniques by comparing the performance of a model trained on raw text data with one trained on processed and TF-IDF weighted text data.

Comparing raw features with processed TF-IDF features on the 20 Newsgroups dataset

In this example, we will simply apply the hashing term frequency transformation to the raw text tokens obtained using a simple whitespace splitting of the document text. We will train a model on this data and evaluate the performance on the test set as we did for the model trained with TF-IDF features:

```
val rawTokens = rdd.map { case (file, text) => text.split(" ") }
val rawTF = texrawTokenst.map(doc => hashingTF.transform(doc))
```

```
val rawTrain = newsgroups.zip(rawTF).map { case (topic, vector) =>
LabeledPoint(newsgroupsMap(topic), vector) }
val rawModel = NaiveBayes.train(rawTrain, lambda = 0.1)
val rawTestTF = testRDD.map { case (file, text) =>
hashingTF.transform(text.split(" ")) }
val rawZippedTest = testLabels.zip(rawTestTF)
val rawTest = rawZippedTest.map { case (topic, vector) =>
LabeledPoint(topic, vector) }
val rawPredictionAndLabel = rawTest.map(p =>
(rawModel.predict(p.features), p.label))
val rawAccuracy = 1.0 * rawPredictionAndLabel.filter(x => x._1 ==
x._2).count() / rawTest.count()
println(rawAccuracy)
val rawMetrics = new MulticlassMetrics(rawPredictionAndLabel)
println(rawMetrics.weightedFMeasure)
```

Perhaps surprisingly, the raw model does quite well, although both accuracy and F-measure are a few percentage points lower than those of the TF-IDF model. This is also partly a reflection of the fact that the naïve Bayes model is well suited to data in the form of raw frequency counts:

0.7661975570897503

0.7628947184990661

Word2Vec models

Until now, we have used a bag-of-words vector, optionally with some weighting scheme such as TF-IDF to represent the text in a document. Another recent class of models that has become popular is related to representing individual words as vectors.

These are generally based in some way on the co-occurrence statistics between the words in a corpus. Once the vector representation is computed, we can use these vectors in ways similar to how we might use TF-IDF vectors (such as using them as features for other machine learning models). One such common use case is computing the similarity between two words with respect to their meanings, based on their vector representations.

Word2Vec refers to a specific implementation of one of these models, often referred to as **distributed vector representations**. The MLlib model uses a **skip-gram** model, which seeks to learn vector representations that take into account the contexts in which words occur.

> While a detailed treatment of Word2Vec is beyond the scope of this book, Spark's documentation at http://spark.apache.org/docs/latest/mllib-feature-extraction.html#word2vec contains some further details on the algorithm as well as links to the reference implementation.
>
> One of the main academic papers underlying Word2Vec is *Tomas Mikolov, Kai Chen, Greg Corrado,* and *Jeffrey Dean. Efficient Estimation of Word Representations in Vector Space. In Proceedings of Workshop at ICLR, 2013.*
>
> It is available at http://arxiv.org/pdf/1301.3781.pdf.
>
> Another recent model in the area of word vector representations is GloVe at http://www-nlp.stanford.edu/projects/glove/.

Word2Vec on the 20 Newsgroups dataset

Training a Word2Vec model in Spark is relatively simple. We will pass in an RDD where each element is a sequence of terms. We can use the RDD of tokenized documents we have already created as input to the model:

```
import org.apache.spark.mllib.feature.Word2Vec
val word2vec = new Word2Vec()
word2vec.setSeed(42)
val word2vecModel = word2vec.fit(tokens)
```

> Note that we used setSeed to set the random seed for model training so that you can see the same results each time the model is trained.

You will see some output similar to the following while the model is being trained:

```
...
14/10/25 14:21:59 INFO Word2Vec: wordCount = 2133172, alpha = 0.0011868763094487506
14/10/25 14:21:59 INFO Word2Vec: wordCount = 2144172, alpha = 0.0010640806039941193
14/10/25 14:21:59 INFO Word2Vec: wordCount = 2155172, alpha = 9.412848985394907E-4
14/10/25 14:21:59 INFO Word2Vec: wordCount = 2166172, alpha = 8.184891930848592E-4
14/10/25 14:22:00 INFO Word2Vec: wordCount = 2177172, alpha = 6.956934876302307E-4
```

Advanced Text Processing with Spark

```
14/10/25 14:22:00 INFO Word2Vec: wordCount = 2188172, alpha =
5.728977821755993E-4
14/10/25 14:22:00 INFO Word2Vec: wordCount = 2199172, alpha =
4.501020767209707E-4
14/10/25 14:22:00 INFO Word2Vec: wordCount = 2210172, alpha =
3.2730637126634213E-4
14/10/25 14:22:01 INFO Word2Vec: wordCount = 2221172, alpha =
2.0451066581171076E-4
14/10/25 14:22:01 INFO Word2Vec: wordCount = 2232172, alpha =
8.171496035708214E-5
...
14/10/25 14:22:02 INFO SparkContext: Job finished: collect at
Word2Vec.scala:368, took 56.585983 s
14/10/25 14:22:02 INFO MappedRDD: Removing RDD 200 from persistence
list
14/10/25 14:22:02 INFO BlockManager: Removing RDD 200
14/10/25 14:22:02 INFO BlockManager: Removing block rdd_200_0
14/10/25 14:22:02 INFO MemoryStore: Block rdd_200_0 of size 9008840
dropped from memory (free 1755596828)
word2vecModel: org.apache.spark.mllib.feature.Word2VecModel =
org.apache.spark.mllib.feature.Word2VecModel@2b94e480
```

Once trained, we can easily find the top 20 synonyms for a given term (that is, the most similar term to the input term, computed by cosine similarity between the word vectors). For example, to find the 20 most similar terms to *hockey*, use the following lines of code:

```
word2vecModel.findSynonyms("hockey", 20).foreach(println)
```

As we can see from the following output, most of the terms relate to hockey or other sports topics:

```
(sport,0.6828256249427795)
(ecac,0.6718048453330994)
(hispanic,0.6519884467124939)
(glens,0.6447514891624451)
(woofers,0.6351765394210815)
(boxscores,0.6009076237678528)
(tournament,0.6006366014480591)
(champs,0.5957855582237244)
(aargh,0.584071934223175)
(playoff,0.5834275484085083)
```

```
(ahl,0.5784651637077332)
(ncaa,0.5680188536643982)
(pool,0.5612311959266663)
(olympic,0.5552600026130676)
(champion,0.5549421310424805)
(filinuk,0.5528956651687622)
(yankees,0.5502706170082092)
(motorcycles,0.5484763979911804)
(calder,0.5481109023094177)
(rec,0.5432182550430298)
```

As another example, we can find 20 synonyms for the term *legislation* as follows:

```
word2vecModel.findSynonyms("legislation", 20).foreach(println)
```

In this case, we observe the terms related to *regulation, politics,* and *business* feature prominently:

```
(accommodates,0.8149217963218689)
(briefed,0.7582570314407349)
(amended,0.7310371994972229)
(telephony,0.7139414548873901)
(aclu,0.7080780863761902)
(pitted,0.7062571048736572)
(licensee,0.6981208324432373)
(agency,0.6880651712417603)
(policies,0.6828961372375488)
(senate,0.6821110844612122)
(businesses,0.6814320087432861)
(permit,0.6797110438346863)
(cpsr,0.6764014959335327)
(cooperation,0.6733141541481018)
(surveillance,0.6670728325843811)
(restricted,0.6666574478149414)
(congress,0.6661365628242493)
(procure,0.6655452251434326)
(industry,0.6650314927101135)
(inquiry,0.6644254922866821)
```

Summary

In this chapter, we took a deeper look into more complex text processing and explored MLlib's text feature extraction capabilities, in particular the TF-IDF term weighting schemes. We covered examples of using the resulting TF-IDF feature vectors to compute document similarity and train a newsgroup topic classification model. Finally, you learned how to use MLlib's cutting-edge Word2Vec model to compute a vector representation of words in a corpus of text and use the trained model to find words with contextual meaning that is similar to a given word.

In the next chapter, we will take a look at online learning, and you will learn how Spark Streaming relates to online learning models.

10
Real-time Machine Learning with Spark Streaming

So far in this book, we have focused on **batch** data processing. That is, all our analysis, feature extraction, and model training has been applied to a fixed set of data that does not change. This fits neatly into Spark's core abstraction of RDDs, which are immutable distributed datasets. Once created, the data underlying the RDD does not change, although we might create new RDDs from the original RDD through Spark's transformation and action operators.

Our attention has also been on batch machine learning models where we train a model on a fixed batch of training data that is usually represented as an RDD of feature vectors (and labels, in the case of supervised learning models).

In this chapter, we will:

- Introduce the concept of online learning, where models are trained and updated on new data as it becomes available
- Explore stream processing using Spark Streaming
- See how Spark Streaming fits together with the online learning approach

Online learning

The batch machine learning methods that we have applied in this book focus on processing an existing fixed set of training data. Typically, these techniques are also iterative, and we have performed multiple passes over our training data in order to converge to an optimal model.

By contrast, online learning is based on performing only one sequential pass through the training data in a fully incremental fashion (that is, one training example at a time). After seeing each training example, the model makes a prediction for this example and then receives the true outcome (for example, the label for classification or real target for regression). The idea behind online learning is that the model continually updates as new information is received instead of being retrained periodically in batch training.

In some settings, when data volume is very large or the process that generates the data is changing rapidly, online learning methods can adapt more quickly and in near real time, without needing to be retrained in an expensive batch process.

However, online learning methods do not have to be used in a purely online manner. In fact, we have already seen an example of using an online learning model in the batch setting when we used **stochastic gradient descent** optimization to train our classification and regression models. SGD updates the model after each training example. However, we still made use of multiple passes over the training data in order to converge to a better result.

In the pure online setting, we do not (or perhaps cannot) make multiple passes over the training data; hence, we need to process each input as it arrives. Online methods also include mini-batch methods where, instead of processing one input at a time, we process a small batch of training data.

Online and batch methods can also be combined in real-world situations. For example, we can periodically retrain our models offline (say, every day) using batch methods. We can then deploy the trained model to production and update it using online methods in real time (that is, during the day, in between batch retraining) to adapt to any changes in the environment.

As we will see in this chapter, the online learning setting can fit neatly into stream processing and the Spark Streaming framework.

 See http://en.wikipedia.org/wiki/Online_machine_learning for more details on online machine learning.

Stream processing

Before covering online learning with Spark, we will first explore the basics of stream processing and introduce the Spark Streaming library.

In addition to the core Spark API and functionality, the Spark project contains another major library (in the same way as MLlib is a major project library) called **Spark Streaming**, which focuses on processing data streams in real time.

A data stream is a continuous sequence of records. Common examples include activity stream data from a web or mobile application, time-stamped log data, transactional data, and event streams from sensor or device networks.

The batch processing approach typically involves saving the data stream to an intermediate storage system (for example, HDFS or a database) and running a batch process on the saved data. In order to generate up-to-date results, the batch process must be run periodically (for example, daily, hourly, or even every few minutes) on the latest data available.

By contrast, the stream-based approach applies processing to the data stream as it is generated. This allows near real-time processing (of the order of a subsecond to a few tenths of a second time frames rather than minutes, hours, days, or even weeks with typical batch processing).

An introduction to Spark Streaming

There are a few different general techniques to deal with stream processing. Two of the most common ones are as follows:

- Treat each record individually and process it as soon as it is seen.
- Combine multiple records into **mini-batches**. These mini-batches can be delineated either by time or by the number of records in a batch.

Spark Streaming takes the second approach. The core primitive in Spark Streaming is the **discretized stream**, or **DStream**. A DStream is a sequence of mini-batches, where each mini-batch is represented as a Spark RDD:

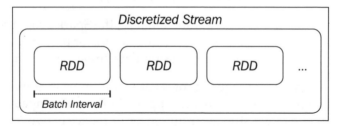

The discretized stream abstraction

A DStream is defined by its input source and a time window called the **batch interval**. The stream is broken up into time periods equal to the batch interval (beginning from the starting time of the application). Each RDD in the stream will contain the records that are received by the Spark Streaming application during a given batch interval. If no data is present in a given interval, the RDD will simply be empty.

Input sources

Spark Streaming **receivers** are responsible for receiving data from an **input source** and converting the raw data into a DStream made up of Spark RDDs.

Spark Streaming supports various input sources, including file-based sources (where the receiver watches for new files arriving at the input location and creates the DStream from the contents read from each new file) and network-based sources (such as receivers that communicate with socket-based sources, the Twitter API stream, Akka actors, or message queues and distributed stream and log transfer frameworks, such as Flume, Kafka, and Amazon Kinesis).

> See the documentation on input sources at http://spark.apache.org/docs/latest/streaming-programming-guide.html#input-dstreams for more details and for links to various advanced sources.

Transformations

As we saw in *Chapter 1, Getting Up and Running with Spark*, and throughout this book, Spark allows us to apply powerful transformations to RDDs. As DStreams are made up of RDDs, Spark Streaming provides a set of transformations available on DStreams; these transformations are similar to those available on RDDs. These include map, flatMap, filter, join, and reduceByKey.

Spark Streaming transformations, such as those applicable to RDDs, operate on each element of a DStream's underlying data. That is, the transformations are effectively applied to each RDD in the DStream, which, in turn, applies the transformation to the elements of the RDD.

Spark Streaming also provides operators such as `reduce` and `count`. These operators return a DStream made up of a single element (for example, the count value for each batch). Unlike the equivalent operators on RDDs, these do not trigger computation on DStreams directly. That is, they are not **actions**, but they are still transformations, as they return another DStream.

Keeping track of state

When we were dealing with batch processing of RDDs, keeping and updating a state variable was relatively straightforward. We could start with a certain state (for example, a count or sum of values) and then use broadcast variables or accumulators to update this state in parallel. Usually, we would then use an RDD action to collect the updated state to the driver and, in turn, update the global state.

With DStreams, this is a little more complex, as we need to keep track of states across batches in a fault-tolerant manner. Conveniently, Spark Streaming provides the `updateStateByKey` function on a DStream of key-value pairs, which takes care of this for us, allowing us to create a stream of arbitrary state information and update it with each batch of data seen. For example, the state could be a global count of the number of times each key has been seen. The state could, thus, represent the number of visits per web page, clicks per advert, tweets per user, or purchases per product, for example.

General transformations

The Spark Streaming API also exposes a general `transform` function that gives us access to the underlying RDD for each batch in the stream. That is, where the higher level functions such as `map` transform a DStream to another DStream, `transform` allows us to apply functions from an RDD to another RDD. For example, we can use the RDD `join` operator to join each batch of the stream to an existing RDD that we computed separately from our streaming application (perhaps, in Spark or some other system).

> The full list of transformations and further information on each of them is provided in the Spark documentation at http://spark.apache.org/docs/latest/streaming-programming-guide.html#transformations-on-dstreams.

Actions

While some of the operators we have seen in Spark Streaming, such as `count`, are not actions as in the batch RDD case, Spark Streaming has the concept of **actions** on DStreams. Actions are **output** operators that, when invoked, trigger computation on the DStream. They are as follows:

- `print`: This prints the first 10 elements of each batch to the console and is typically used for debugging and testing.
- `saveAsObjectFile`, `saveAsTextFiles`, and `saveAsHadoopFiles`: These functions output each batch to a Hadoop-compatible filesystem with a filename (if applicable) derived from the batch start timestamp.
- `forEachRDD`: This operator is the most generic and allows us to apply any arbitrary processing to the RDDs within each batch of a DStream. It is used to apply *side effects*, such as saving data to an external system, printing it for testing, exporting it to a dashboard, and so on.

> Note that like batch processing with Spark, DStream operators are **lazy**. In the same way in which we need to call an action, such as count, on an RDD to ensure that processing takes place, we need to call one of the preceding action operators in order to trigger computation on a DStream. Otherwise, our streaming application will not actually perform any computation.

Window operators

As Spark Streaming operates on time-ordered batched streams of data, it introduces a new concept, which is that of **windowing**. A `window` function computes a transformation over a sliding window applied to the stream.

A window is defined by the length of the window and the sliding interval. For example, with a 10-second window and a 5-second sliding interval, we will compute results every 5 seconds, based on the latest 10 seconds of data in the DStream. For example, we might wish to calculate the top websites by page view numbers over the last 10 seconds and recompute this metric every 5 seconds using a sliding window.

The following figure illustrates a windowed DStream:

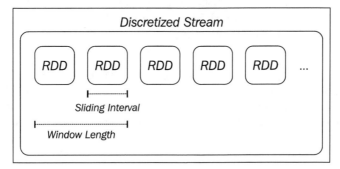

A windowed DStream

Caching and fault tolerance with Spark Streaming

Like Spark RDDs, DStreams can be cached in memory. The use cases for caching are similar to those for RDDs—if we expect to access the data in a DStream multiple times (perhaps performing multiple types of analysis or aggregation or outputting to multiple external systems), we will benefit from caching the data. Stateful operators, which include `window` functions and `updateStateByKey`, do this automatically for efficiency.

Recall that RDDs are immutable datasets and are defined by their input data source and **lineage**—that is, the set of transformations and actions that are applied to the RDD. Fault tolerance in RDDs works by recreating the RDD (or partition of an RDD) that is lost due to the failure of a worker node.

As DStreams are themselves batches of RDDs, they can also be recomputed as required to deal with worker node failure. However, this depends on the input data still being available. If the data source itself is fault-tolerant and persistent (such as HDFS or some other fault-tolerant data store), then the DStream can be recomputed.

If data stream sources are delivered over a network (which is a common case with stream processing), Spark Streaming's default persistence behavior is to replicate data to two worker nodes. This allows network DStreams to be recomputed in the case of failure. Note, however, that any data received by a node but *not yet replicated* might be lost when a node fails.

Spark Streaming also supports recovery of the driver node in the event of failure. However, currently, for network-based sources, data in the memory of worker nodes will be lost in this case. Hence, Spark Streaming is not fully fault-tolerant in the face of failure of the driver node or application.

> See http://spark.apache.org/docs/latest/streaming-programming-guide.html#caching—persistence and http://spark.apache.org/docs/latest/streaming-programming-guide.html#fault-tolerance-properties for more details.

Creating a Spark Streaming application

We will now work through creating our first Spark Streaming application to illustrate some of the basic concepts around Spark Streaming that we introduced earlier.

We will expand on the example applications used in *Chapter 1*, *Getting Up and Running with Spark*, where we used a small example dataset of product purchase events. For this example, instead of using a static set of data, we will create a simple producer application that will randomly generate events and send them over a network connection. We will then create a few Spark Streaming consumer applications that will process this event stream.

The sample project for this chapter contains the code you will need. It is called scala-spark-streaming-app. It consists of a Scala SBT project definition file, the example application source code, and a \src\main\resources directory that contains a file called names.csv.

The build.sbt file for the project contains the following project definition:

```
name := "scala-spark-streaming-app"

version := "1.0"

scalaVersion := "2.10.4"

libraryDependencies += "org.apache.spark" %% "spark-mllib"
% "1.1.0"

libraryDependencies += "org.apache.spark" %% "spark-streaming"
% "1.1.0"
```

Note that we added a dependency on Spark MLlib and Spark Streaming, which includes the dependency on the Spark core.

The `names.csv` file contains a set of 20 randomly generated user names. We will use these names as part of our data generation function in our producer application:

Miguel,Eric,James,Juan,Shawn,James,Doug,Gary,Frank,Janet,Michael, James,Malinda,Mike,Elaine,Kevin,Janet,Richard,Saul,Manuela

The producer application

Our producer needs to create a network connection and generate some random purchase event data to send over this connection. First, we will define our object and main method definition. We will then read the random names from the `names.csv` resource and create a set of products with prices, from which we will generate our random product events:

```
/**
 * A producer application that generates random "product events",
up to 5 per second, and sends them over a
 * network connection
 */
object StreamingProducer {

  def main(args: Array[String]) {

    val random = new Random()

    // Maximum number of events per second
    val MaxEvents = 6

    // Read the list of possible names
    val namesResource =
    this.getClass.getResourceAsStream("/names.csv")
    val names = scala.io.Source.fromInputStream(namesResource)
      .getLines()
      .toList
      .head
      .split(",")
      .toSeq

    // Generate a sequence of possible products
    val products = Seq(
```

```
    "iPhone Cover" -> 9.99,
    "Headphones" -> 5.49,
    "Samsung Galaxy Cover" -> 8.95,
    "iPad Cover" -> 7.49
)
```

Using the list of names and map of product name to price, we will create a function that will randomly pick a product and name from these sources, generating a specified number of product events:

```
/** Generate a number of random product events */
def generateProductEvents(n: Int) = {
  (1 to n).map { i =>
    val (product, price) =
    products(random.nextInt(products.size))
    val user = random.shuffle(names).head
    (user, product, price)
  }
}
```

Finally, we will create a network socket and set our producer to listen on this socket. As soon as a connection is made (which will come from our consumer streaming application), the producer will start generating random events at a random rate between 0 and 5 per second:

```
// create a network producer
val listener = new ServerSocket(9999)
println("Listening on port: 9999")

while (true) {
  val socket = listener.accept()
  new Thread() {
    override def run = {
      println("Got client connected from: " +
      socket.getInetAddress)
      val out = new PrintWriter(socket.getOutputStream(),
      true)

      while (true) {
        Thread.sleep(1000)
        val num = random.nextInt(MaxEvents)
        val productEvents = generateProductEvents(num)
        productEvents.foreach{ event =>
```

```
              out.write(event.productIterator.mkString(","))
              out.write("\n")
            }
            out.flush()
            println(s"Created $num events...")
          }
          socket.close()
        }
      }.start()
    }
  }
}
```

 This producer example is based on the `PageViewGenerator` example in the Spark Streaming examples.

The producer can be run by changing into the base directory of `scala-spark-streaming-app` and using SBT to run the application, as we did in *Chapter 1, Getting Up and Running with Spark*:

```
>cd scala-spark-streaming-app
>sbt
[info] ...
>
```

Use the `run` command to execute the application:

```
>run
```

You should see output similar to the following:

```
...
Multiple main classes detected, select one to run:

 [1] StreamingProducer
 [2] SimpleStreamingApp
 [3] StreamingAnalyticsApp
 [4] StreamingStateApp
 [5] StreamingModelProducer
 [6] SimpleStreamingModel
 [7] MonitoringStreamingModel

Enter number:
```

Select the `StreamingProducer` option. The application will start running, and you should see the following output:

```
[info] Running StreamingProducer
Listening on port: 9999
```

We can see that the producer is listening on port `9999`, waiting for our consumer application to connect.

Creating a basic streaming application

Next, we will create our first streaming program. We will simply connect to the producer and print out the contents of each batch. Our streaming code looks like this:

```scala
/**
 * A simple Spark Streaming app in Scala
 */
object SimpleStreamingApp {

  def main(args: Array[String]) {

    val ssc = new StreamingContext("local[2]",
    "First Streaming App", Seconds(10))
    val stream = ssc.socketTextStream("localhost", 9999)

    // here we simply print out the first few elements of each
    // batch
    stream.print()
    ssc.start()
    ssc.awaitTermination()

  }
}
```

It looks fairly simple, and it is mostly due to the fact that Spark Streaming takes care of all the complexity for us. First, we initialized a `StreamingContext` (which is the streaming equivalent of the `SparkContext` we have used so far), specifying similar configuration options that are used to create a `SparkContext`. Notice, however, that here we are required to provide the batch interval, which we set to 10 seconds.

We then created our data stream using a predefined streaming source, `socketTextStream`, which reads text from a socket host and port and creates a `DStream[String]`. We then called the `print` function on the DStream; this function prints out the first few elements of each batch.

[Calling `print` on a DStream is similar to calling `take` on an RDD. It displays only the first few elements.]

We can run this program using SBT. Open a second terminal window, leaving the producer program running, and run `sbt`:

```
>sbt
[info] ...
>run
....
```

Again, you should see a few options to select:

```
Multiple main classes detected, select one to run:

 [1] StreamingProducer
 [2] SimpleStreamingApp
 [3] StreamingAnalyticsApp
 [4] StreamingStateApp
 [5] StreamingModelProducer
 [6] SimpleStreamingModel
 [7] MonitoringStreamingModel
```

Run the `SimpleStreamingApp` main class. You should see the streaming program start up, displaying output similar to the one shown here:

```
...
14/11/15 21:02:23 INFO scheduler.ReceiverTracker: ReceiverTracker started
14/11/15 21:02:23 INFO dstream.ForEachDStream: metadataCleanupDelay = -1
14/11/15 21:02:23 INFO dstream.SocketInputDStream: metadataCleanupDelay = -1
14/11/15 21:02:23 INFO dstream.SocketInputDStream: Slide time = 10000 ms
14/11/15 21:02:23 INFO dstream.SocketInputDStream: Storage level = StorageLevel(false, false, false, false, 1)
14/11/15 21:02:23 INFO dstream.SocketInputDStream: Checkpoint interval = null
```

```
14/11/15 21:02:23 INFO dstream.SocketInputDStream: Remember duration
= 10000 ms
14/11/15 21:02:23 INFO dstream.SocketInputDStream: Initialized and
validated org.apache.spark.streaming.dstream.SocketInputDStream@ff3436d
14/11/15 21:02:23 INFO dstream.ForEachDStream: Slide time = 10000 ms
14/11/15 21:02:23 INFO dstream.ForEachDStream: Storage level =
StorageLevel(false, false, false, false, 1)
14/11/15 21:02:23 INFO dstream.ForEachDStream: Checkpoint interval =
null
14/11/15 21:02:23 INFO dstream.ForEachDStream: Remember duration =
10000 ms
14/11/15 21:02:23 INFO dstream.ForEachDStream: Initialized and
validated org.apache.spark.streaming.dstream.ForEachDStream@5a10b6e8
14/11/15 21:02:23 INFO scheduler.ReceiverTracker: Starting 1
receivers
14/11/15 21:02:23 INFO spark.SparkContext: Starting job: runJob at
ReceiverTracker.scala:275
...
```

At the same time, you should see that the terminal window running the producer displays something like the following:

```
...
Got client connected from: /127.0.0.1
Created 2 events...
Created 2 events...
Created 3 events...
Created 1 events...
Created 5 events...
...
```

After about 10 seconds, which is the time of our streaming batch interval, Spark Streaming will trigger a computation on the stream due to our use of the `print` operator. This should display the first few events in the batch, which will look something like the following output:

```
...
14/11/15 21:02:30 INFO spark.SparkContext: Job finished: take at
DStream.scala:608, took 0.05596 s
-------------------------------------------
Time: 1416078150000 ms
```

```
------------------------------------------
Michael,Headphones,5.49
Frank,Samsung Galaxy Cover,8.95
Eric,Headphones,5.49
Malinda,iPad Cover,7.49
James,iPhone Cover,9.99
James,Headphones,5.49
Doug,iPhone Cover,9.99
Juan,Headphones,5.49
James,iPhone Cover,9.99
Richard,iPad Cover,7.49
...
```

> Note that you might see different results, as the producer generates a random number of random events each second.

You can terminate the streaming app by pressing *Ctrl* + *C*. If you want to, you can also terminate the producer (if you do, you will need to restart it again before starting the next streaming programs that we will create).

Streaming analytics

Next, we will create a slightly more complex streaming program. In *Chapter 1, Getting Up and Running with Spark*, we calculated a few metrics on our dataset of product purchases. These included the total number of purchases, the number of unique users, the total revenue, and the most popular product (together with its number of purchases and total revenue).

In this example, we will compute the same metrics on our stream of purchase events. The key difference is that these metrics will be computed per batch and printed out.

We will define our streaming application code here:

```
/**
 * A more complex Streaming app, which computes statistics and
 prints the results for each batch in a DStream
 */
object StreamingAnalyticsApp {

  def main(args: Array[String]) {
```

```
    val ssc = new StreamingContext("local[2]",
"First Streaming App", Seconds(10))
    val stream = ssc.socketTextStream("localhost", 9999)

    // create stream of events from raw text elements
    val events = stream.map { record =>
      val event = record.split(",")
      (event(0), event(1), event(2))
    }
```

First, we created exactly the same `StreamingContext` and socket stream as we did earlier. Our next step is to apply a `map` transformation to the raw text, where each record is a comma-separated string representing the purchase event. The `map` function splits the text and creates a tuple of `(user, product, price)`. This illustrates the use of `map` on a DStream and how it is the same as if we had been operating on an RDD.

Next, we will use `foreachRDD` to apply arbitrary processing on each RDD in the stream to compute our desired metrics and print them to the console:

```
    /*
      We compute and print out stats for each batch.
      Since each batch is an RDD, we call forEeachRDD on the
      DStream, and apply the usual RDD functions
      we used in Chapter 1.
    */
    events.foreachRDD { (rdd, time) =>
      val numPurchases = rdd.count()
      val uniqueUsers = rdd.map { case (user, _, _) => user
      }.distinct().count()
      val totalRevenue = rdd.map { case (_, _, price) =>
      price.toDouble }.sum()
      val productsByPopularity = rdd
          .map { case (user, product, price) => (product, 1) }
          .reduceByKey(_ + _)
          .collect()
          .sortBy(-_._2)
      val mostPopular = productsByPopularity(0)

      val formatter = new SimpleDateFormat
      val dateStr = formatter.format(new Date(time.milliseconds))
      println(s"== Batch start time: $dateStr ==")
```

```
        println("Total purchases: " + numPurchases)
        println("Unique users: " + uniqueUsers)
        println("Total revenue: " + totalRevenue)
        println("Most popular product: %s with %d
        purchases".format(mostPopular._1, mostPopular._2))
    }

    // start the context
    ssc.start()
    ssc.awaitTermination()

  }

}
```

If you compare the code operating on the RDDs inside the preceding `foreachRDD` block with that used in *Chapter 1, Getting Up and Running with Spark,* you will notice that it is virtually the same code. This shows that we can apply any RDD-related processing we wish within the streaming setting by operating on the underlying RDDs, as well as using the built-in higher level streaming operations.

Let's run the streaming program again by calling `sbt run` and selecting `StreamingAnalyticsApp`.

> Remember that you might also need to restart the producer if you previously terminated the program. This should be done before starting the streaming application.

After about 10 seconds, you should see output from the streaming program similar to the following:

...

14/11/15 21:27:30 INFO spark.SparkContext: Job finished: collect at Streaming.scala:125, took 0.071145 s

== Batch start time: 2014/11/15 9:27 PM ==

Total purchases: 16

Unique users: 10

Total revenue: 123.72

Most popular product: iPad Cover with 6 purchases

...

You can again terminate the streaming program using *Ctrl + C*.

Stateful streaming

As a final example, we will apply the concept of **stateful** streaming using the `updateStateByKey` function to compute a global state of revenue and number of purchases per user, which will be updated with new data from each 10-second batch. Our `StreamingStateApp` app is shown here:

```
object StreamingStateApp {
  import org.apache.spark.streaming.StreamingContext._
```

We will first define an `updateState` function that will compute the new state from the running state value and the new data in the current batch. Our state, in this case, is a tuple of `(number of products, revenue)` pairs, which we will keep for each user. We will compute the new state given the set of `(product, revenue)` pairs for the current batch and the accumulated state at the current time.

Notice that we will deal with an `Option` value for the current state, as it might be empty (which will be the case for the first batch), and we need to define a default value, which we will do using `getOrElse` as shown here:

```
def updateState(prices: Seq[(String, Double)], currentTotal:
Option[(Int, Double)]) = {
  val currentRevenue = prices.map(_._2).sum
  val currentNumberPurchases = prices.size
  val state = currentTotal.getOrElse((0, 0.0))
  Some((currentNumberPurchases + state._1, currentRevenue +
  state._2))
}

def main(args: Array[String]) {

  val ssc = new StreamingContext("local[2]", "First Streaming
  App", Seconds(10))
  // for stateful operations, we need to set a checkpoint
  // location
  ssc.checkpoint("/tmp/sparkstreaming/")
  val stream = ssc.socketTextStream("localhost", 9999)

  // create stream of events from raw text elements
  val events = stream.map { record =>
    val event = record.split(",")
    (event(0), event(1), event(2).toDouble)
  }
```

```
        val users = events.map{ case (user, product, price) => (user,
        (product, price)) }
        val revenuePerUser = users.updateStateByKey(updateState)
        revenuePerUser.print()

        // start the context
        ssc.start()
        ssc.awaitTermination()

    }
}
```

After applying the same string split transformation we used in our previous example, we called `updateStateByKey` on our DStream, passing in our defined `updateState` function. We then printed the results to the console.

Start the streaming example using `sbt run` and by selecting [4] `StreamingStateApp` (also restart the producer program if necessary).

After around 10 seconds, you will start to see the first set of state output. We will wait another 10 seconds to see the next set of output. You will see the overall global state being updated:

```
...
-------------------------------------------
Time: 1416080440000 ms
-------------------------------------------
(Janet,(2,10.98))
(Frank,(1,5.49))
(James,(2,12.98))
(Malinda,(1,9.99))
(Elaine,(3,29.97))
(Gary,(2,12.98))
(Miguel,(3,20.47))
(Saul,(1,5.49))
(Manuela,(2,18.939999999999998))
(Eric,(2,18.939999999999998))
...
```

```
-------------------------------------------
Time: 1416080441000 ms
-------------------------------------------
(Janet,(6,34.94))
(Juan,(4,33.92))
(Frank,(2,14.44))
(James,(7,48.93000000000001))
(Malinda,(1,9.99))
(Elaine,(7,61.89))
(Gary,(4,28.46))
(Michael,(1,8.95))
(Richard,(2,16.439999999999998))
(Miguel,(5,35.95))
...
```

We can see that the number of purchases and revenue totals for each user are added to with each batch of data.

> Now, see if you can adapt this example to use Spark Streaming's window functions. For example, you can compute similar statistics per user over the past minute, sliding every 30 seconds.

Online learning with Spark Streaming

As we have seen, Spark Streaming makes it easy to work with data streams in a way that should be familiar to us from working with RDDs. Using Spark's stream processing primitives combined with the online learning capabilities of MLlib's SGD-based methods, we can create real-time machine learning models that we can update on new data in the stream as it arrives.

Streaming regression

Spark provides a built-in streaming machine learning model in the `StreamingLinearAlgorithm` class. Currently, only a linear regression implementation is available – `StreamingLinearRegressionWithSGD` – but future versions will include classification.

The streaming regression model provides two methods for usage:

- `trainOn`: This takes `DStream[LabeledPoint]` as its argument. This tells the model to train on every batch in the input DStream. It can be called multiple times to train on different streams.
- `predictOn`: This also takes `DStream[LabeledPoint]`. This tells the model to make predictions on the input DStream, returning a new `DStream[Double]` that contains the model predictions.

Under the hood, the streaming regression model uses `foreachRDD` and `map` to accomplish this. It also updates the model variable after each batch and exposes the latest trained model, which allows us to use this model in other applications or save it to an external location.

The streaming regression model can be configured with parameters for step size and number of iterations in the same way as standard batch regression—the model class used is the same. We can also set the initial model weight vector.

When we first start training a model, we can set the initial weights to a zero vector, or a random vector, or perhaps load the latest model from the result of an offline batch process. We can also decide to save the model periodically to an external system and use the latest model state as the starting point (for example, in the case of a restart after a node or application failure).

A simple streaming regression program

To illustrate the use of streaming regression, we will create a simple example similar to the preceding one, which uses simulated data. We will write a producer program that generates random feature vectors and target variables, given a fixed, known weight vector, and writes each training example to a network stream.

Our consumer application will run a streaming regression model, training and then testing on our simulated data stream. Our first example consumer will simply print its predictions to the console.

Creating a streaming data producer

The data producer operates in a manner similar to our product event producer example. Recall from *Chapter 5, Building a Classification Model with Spark*, that a linear model is a linear combination (or vector dot product) of a weight vector, w, and a feature vector, x (that is, wTx). Our producer will generate synthetic data using a fixed, known weight vector and randomly generated feature vectors. This data fits the linear model formulation exactly, so we will expect our regression model to learn the true weight vector fairly easily.

First, we will set up a maximum number of events per second (say, 100) and the number of features in our feature vector (also 100 in this example):

```
/**
 * A producer application that generates random linear regression
data.
 */
object StreamingModelProducer {
  import breeze.linalg._

  def main(args: Array[String]) {

    // Maximum number of events per second
    val MaxEvents = 100
    val NumFeatures = 100

    val random = new Random()
```

The `generateRandomArray` function creates an array of the specified size where the entries are randomly generated from a normal distribution. We will use this function initially to generate our known weight vector, w, which will be fixed throughout the life of the producer. We will also create a random `intercept` value that will also be fixed. The weight vector and `intercept` will be used to generate each data point in our stream:

```
    /** Function to generate a normally distributed dense vector
     */
    def generateRandomArray(n: Int) = Array.tabulate(n)( _ =>
random.nextGaussian())

    // Generate a fixed random model weight vector
    val w = new DenseVector(generateRandomArray(NumFeatures))
    val intercept = random.nextGaussian() * 10
```

We will also need a function to generate a specified number of random data points. Each event is made up of a random feature vector and the target that we get from computing the dot product of our known weight vector with the random feature vector and adding the `intercept` value:

```
    /** Generate a number of random data events*/
    def generateNoisyData(n: Int) = {
      (1 to n).map { i =>
        val x = new DenseVector(generateRandomArray(NumFeatures))
        val y: Double = w.dot(x)
```

```
        val noisy = y + intercept
        (noisy, x)
      }
    }
```

Finally, we will use code similar to our previous producer to instantiate a network connection and send a random number of data points (between 0 and 100) in text format over the network each second:

```
    // create a network producer
    val listener = new ServerSocket(9999)
    println("Listening on port: 9999")

    while (true) {
      val socket = listener.accept()
      new Thread() {
        override def run = {
          println("Got client connected from: " +
          socket.getInetAddress)
          val out = new PrintWriter(socket.getOutputStream(),
          true)

          while (true) {
            Thread.sleep(1000)
            val num = random.nextInt(MaxEvents)
            val data = generateNoisyData(num)
            data.foreach { case (y, x) =>
              val xStr = x.data.mkString(",")
              val eventStr = s"$y\t$xStr"
              out.write(eventStr)
              out.write("\n")
            }
            out.flush()
            println(s"Created $num events...")
          }
          socket.close()
        }
      }.start()
    }
  }
}
```

You can start the producer using `sbt run`, followed by choosing to execute the `StreamingModelProducer` main method. This should result in the following output, thus indicating that the producer program is waiting for connections from our streaming regression application:

```
[info] Running StreamingModelProducer
Listening on port: 9999
```

Creating a streaming regression model

In the next step in our example, we will create a streaming regression program. The basic layout and setup is the same as our previous streaming analytics examples:

```
/**
 * A simple streaming linear regression that prints out predicted
value for each batch
 */
object SimpleStreamingModel {

  def main(args: Array[String]) {

    val ssc = new StreamingContext("local[2]", "First Streaming App", Seconds(10))
    val stream = ssc.socketTextStream("localhost", 9999)
```

Here, we will set up the number of features to match the records in our input data stream. We will then create a zero vector to use as the initial weight vector of our streaming regression model. Finally, we will select the number of iterations and step size:

```
val NumFeatures = 100
    val zeroVector = DenseVector.zeros[Double](NumFeatures)
    val model = new StreamingLinearRegressionWithSGD()
      .setInitialWeights(Vectors.dense(zeroVector.data))
      .setNumIterations(1)
      .setStepSize(0.01)
```

Next, we will again use the `map` function to transform the input DStream, where each record is a string representation of our input data, into a `LabeledPoint` instance that contains the target value and feature vector:

```
    // create a stream of labeled points
    val labeledStream = stream.map { event =>
      val split = event.split("\t")
```

```
            val y = split(0).toDouble
            val features = split(1).split(",").map(_.toDouble)
            LabeledPoint(label = y, features = Vectors.dense(features))
        }
```

The final step is to tell the model to train and test on our transformed DStream and also to print out the first few elements of each batch in the DStream of predicted values:

```
        // train and test model on the stream, and print predictions
        // for illustrative purposes
        model.trainOn(labeledStream)
        model.predictOn(labeledStream).print()

        ssc.start()
        ssc.awaitTermination()

    }
}
```

> Note that because we are using the same MLlib model classes for streaming as we did for batch processing, we can, if we choose, perform multiple iterations over the training data in each batch (which is just an RDD of `LabeledPoint` instances).
>
> Here, we will set the number of iterations to 1 to simulate purely online learning. In practice, you can set the number of iterations higher, but note that the training time per batch will go up. If the training time per batch is much higher than the batch interval, the streaming model will start to lag behind the velocity of the data stream.
>
> This can be handled by decreasing the number of iterations, increasing the batch interval, or increasing the parallelism of our streaming program by adding more Spark workers.

Now, we're ready to run `SimpleStreamingModel` in our second terminal window using `sbt run` in the same way as we did for the producer (remember to select the correct main method for SBT to execute). Once the streaming program starts running, you should see the following output in the producer console:

```
Got client connected from: /127.0.0.1
...
Created 10 events...
Created 83 events...
Created 75 events...
...
```

Real-time Machine Learning with Spark Streaming

After about 10 seconds, you should start seeing the model predictions being printed to the streaming application console, similar to those shown here:

```
14/11/16 14:54:00 INFO StreamingLinearRegressionWithSGD: Model
updated at time 1416142440000 ms
```

```
14/11/16 14:54:00 INFO StreamingLinearRegressionWithSGD: Current
model: weights, [0.05160959387864821,0.05122747155689144,-
0.17224086785756998,0.05822993392274008,0.07848094246845688,-
0.1298315806501979,0.006059323642394124, ...
```

...

```
14/11/16 14:54:00 INFO JobScheduler: Finished job streaming job
1416142440000 ms.0 from job set of time 1416142440000 ms
```

```
14/11/16 14:54:00 INFO JobScheduler: Starting job streaming job
1416142440000 ms.1 from job set of time 1416142440000 ms
```

```
14/11/16 14:54:00 INFO SparkContext: Starting job: take at
DStream.scala:608
```

```
14/11/16 14:54:00 INFO DAGScheduler: Got job 3 (take at
DStream.scala:608) with 1 output partitions (allowLocal=true)
```

```
14/11/16 14:54:00 INFO DAGScheduler: Final stage: Stage 3(take at
DStream.scala:608)
```

```
14/11/16 14:54:00 INFO DAGScheduler: Parents of final stage: List()
```

```
14/11/16 14:54:00 INFO DAGScheduler: Missing parents: List()
```

```
14/11/16 14:54:00 INFO DAGScheduler: Computing the requested
partition locally
```

```
14/11/16 14:54:00 INFO SparkContext: Job finished: take at
DStream.scala:608, took 0.014064 s
```

```
-------------------------------------------
Time: 1416142440000 ms
-------------------------------------------
-2.0851430248312526
4.609405228401022
2.817934589675725
3.3526557917118813
4.624236379848475
-2.3509098272485156
-0.7228551577759544
2.914231548990703
0.896926579927631
1.1968162940541283
...
```

Congratulations! You've created your first streaming online learning model!

You can shut down the streaming application (and, optionally, the producer) by pressing *Ctrl + C* in each terminal window.

Streaming K-means

MLlib also includes a streaming version of K-means clustering; this is called `StreamingKMeans`. This model is an extension of the mini-batch K-means algorithm where the model is updated with each batch based on a combination between the cluster centers computed from the previous batches and the cluster centers computed for the current batch.

`StreamingKMeans` supports a *forgetfulness* parameter *alpha* (set using the `setDecayFactor` method); this controls how aggressive the model is in giving weight to newer data. An alpha value of 0 means the model will only use new data, while with an alpha value of 1, all data since the beginning of the streaming application will be used.

We will not cover streaming K-means further here (the Spark documentation at http://spark.apache.org/docs/latest/mllib-clustering.html#streaming-clustering contains further detail and an example). However, perhaps you could try to adapt the preceding streaming regression data producer to generate input data for a `StreamingKMeans` model. You could also adapt the streaming regression application to use `StreamingKMeans`.

You can create the clustering data producer by first selecting a number of clusters, *K*, and then generating each data point by:

- Randomly selecting a cluster index.
- Generating a random vector using specific normal distribution parameters for each cluster. That is, each of the *K* clusters will have a mean and variance parameter, from which the random vectors will be generated using an approach similar to our preceding `generateRandomArray` function.

In this way, each data point that belongs to the same cluster will be drawn from the same distribution, so our streaming clustering model should be able to learn the correct cluster centers over time.

Online model evaluation

Combining machine learning with Spark Streaming has many potential applications and use cases, including keeping a model or set of models up to date on new training data as it arrives, thus enabling them to adapt quickly to changing situations or contexts.

Another useful application is to track and compare the performance of multiple models in an online manner and, possibly, also perform model selection in real time so that the best performing model is always used to generate predictions on live data.

This can be used to do real-time "A/B testing" of models, or combined with more advanced online selection and learning techniques, such as Bayesian update approaches and bandit algorithms. It can also be used simply to monitor model performance in real time, thus being able to respond or adapt if performance degrades for some reason.

In this section, we will walk through a simple extension to our streaming regression example. In this example, we will compare the evolving error rate of two models with different parameters as they see more and more data in our input stream.

Comparing model performance with Spark Streaming

As we have used a known weight vector and intercept to generate the training data in our producer application, we would expect our model to eventually learn this underlying weight vector (in the absence of random noise, which we do not add for this example).

Therefore, we should see the model's error rate decrease over time, as it sees more and more data. We can also use standard regression error metrics to compare the performance of multiple models.

In this example, we will create two models with different learning rates, training them both on the same data stream. We will then make predictions for each model and measure the **mean-squared error** (MSE) and **root mean-squared error** (RMSE) metrics for each batch.

Our new monitored streaming model code is shown here:

```
/**
 * A streaming regression model that compares the model
performance of two models, printing out metrics for
```

```
 * each batch
 */
object MonitoringStreamingModel {
  import org.apache.spark.SparkContext._

  def main(args: Array[String]) {

    val ssc = new StreamingContext("local[2]", "First Streaming
    App", Seconds(10))
    val stream = ssc.socketTextStream("localhost", 9999)

    val NumFeatures = 100
    val zeroVector = DenseVector.zeros[Double](NumFeatures)
    val model1 = new StreamingLinearRegressionWithSGD()
      .setInitialWeights(Vectors.dense(zeroVector.data))
      .setNumIterations(1)
      .setStepSize(0.01)

    val model2 = new StreamingLinearRegressionWithSGD()
      .setInitialWeights(Vectors.dense(zeroVector.data))
      .setNumIterations(1)
      .setStepSize(1.0)
// create a stream of labeled points
    val labeledStream = stream.map { event =>
      val split = event.split("\t")
      val y = split(0).toDouble
      val features = split(1).split(",").map(_.toDouble)
      LabeledPoint(label = y, features = Vectors.dense(features))
    }
```

Note that most of the preceding setup code is the same as our simple streaming model example. However, we created two instances of StreamingLinearRegressionWithSGD: one with a learning rate of 0.01 and one with the learning rate set to 1.0.

Next, we will train each model on our input stream, and using Spark Streaming's transform function, we will create a new DStream that contains the error rates for each model:

```
// train both models on the same stream
model1.trainOn(labeledStream)
model2.trainOn(labeledStream)
```

```
    // use transform to create a stream with model error rates
    val predsAndTrue = labeledStream.transform { rdd =>
      val latest1 = model1.latestModel()
      val latest2 = model2.latestModel()
      rdd.map { point =>
        val pred1 = latest1.predict(point.features)
        val pred2 = latest2.predict(point.features)
        (pred1 - point.label, pred2 - point.label)
      }
    }
```

Finally, we will use `foreachRDD` to compute the MSE and RMSE metrics for each model and print them to the console:

```
    // print out the MSE and RMSE metrics for each model per batch
    predsAndTrue.foreachRDD { (rdd, time) =>
      val mse1 = rdd.map { case (err1, err2) => err1 * err1
      }.mean()
      val rmse1 = math.sqrt(mse1)
      val mse2 = rdd.map { case (err1, err2) => err2 * err2
      }.mean()
      val rmse2 = math.sqrt(mse2)
      println(
        s"""
           |-------------------------------------------
           |Time: $time
           |-------------------------------------------
         """.stripMargin)
      println(s"MSE current batch: Model 1: $mse1; Model 2: $mse2")
      println(s"RMSE current batch: Model 1: $rmse1; Model 2: $rmse2")
      println("...\n")
    }

    ssc.start()
    ssc.awaitTermination()

  }
}
```

If you terminated the producer earlier, start it again by executing `sbt run` and selecting `StreamingModelProducer`. Once the producer is running again, in your second terminal window, execute `sbt run` and choose the main class for `MonitoringStreamingModel`.

You should see the streaming program startup, and after about 10 seconds, the first batch will be processed, printing output similar to the following:

```
...
14/11/16 14:56:11 INFO SparkContext: Job finished: mean at
StreamingModel.scala:159, took 0.09122 s


-------------------------------------------
Time: 1416142570000 ms
-------------------------------------------


MSE current batch: Model 1: 97.9475827857361; Model 2:
97.9475827857361
RMSE current batch: Model 1: 9.896847113385965; Model 2:
9.896847113385965
...
```

Since both models start with the same initial weight vector, we see that they both make the same predictions on this first batch and, therefore, have the same error.

If we leave the streaming program running for a few minutes, we should eventually see that one of the models has started converging, leading to a lower and lower error, while the other model has tended to diverge to a poorer model due to the overly high learning rate:

```
...
14/11/16 14:57:30 INFO SparkContext: Job finished: mean at
StreamingModel.scala:159, took 0.069175 s


-------------------------------------------
Time: 1416142650000 ms
-------------------------------------------


MSE current batch: Model 1: 75.54543031658632; Model 2:
10318.213926882852
RMSE current batch: Model 1: 8.691687426304878; Model 2:
101.57860959317593
...
```

If you leave the program running for a number of minutes, you should eventually see the first model's error rate getting quite small:

```
...
14/11/16 17:27:00 INFO SparkContext: Job finished: mean at
StreamingModel.scala:159, took 0.037856 s


-------------------------------------------
Time: 1416151620000 ms
-------------------------------------------


MSE current batch: Model 1: 6.551475362521364; Model 2:
1.057088005456417E26
RMSE current batch: Model 1: 2.559584998104451; Model 2:
1.0281478519436867E13
...
```

 Note again that due to random data generation, you might see different results, but the overall result should be the same — in the first batch, the models will have the same error, and subsequently, the first model should start to generate to a smaller and smaller error.

Summary

In this chapter, we connected some of the dots between online machine learning and streaming data analysis. We introduced the Spark Streaming library and API for continuous processing of data streams based on familiar RDD functionality and worked through examples of streaming analytics applications that illustrate this functionality.

Finally, we used MLlib's streaming regression model in a streaming application that involves computing and comparing model performance on a stream of input feature vectors.

Index

Symbols

1-of-k encoding 71
20 Newsgroups dataset
 about 251
 document similarity, used with 268-270
 exploring 253, 254
 text classifier, training on 271, 272
 TF-IDF features, extracting from 251-253
 URL 251
 Word2Vec models, used on 275-277

A

Abstract Window Toolkit (AWT) 229
accumulators 19, 20
additive smoothing
 URL 156
agglomerative clustering 203
alpha parameter 98
Alternating Least Squares (ALS) 91
Amazon AWS public datasets
 about 52
 URL 52
Amazon EC2
 EC2 Spark cluster, launching 31-35
 Spark, running on 30, 31
Amazon Web Services account
 URL 30
Anaconda
 URL 56
analytics
 streaming 293-295

architecture, machine learning system 48, 49
AUC, classification models 138-140
AWS console
 URL 30

B

bad data
 filling 69
bag-of-words model 248
base form 264
basic streaming application
 creating 290-293
batch data processing 279
batch interval 282
bike sharing dataset
 features, extracting from 164-167
 performance metrics, computing on 175
 regression model, training on 171-173
binary classification 117
Breeze library 212
broadcast variable 19, 20
built-in evaluation functions
 MAP 113
 MSE 113
 RMSE 113
 using 113
business use cases, machine learning system
 about 39
 customer segmentation 40
 personalization 40
 predictive modeling and analytics 41
 targeted marketing 40

C

categorical features
  about 71, 72
  timestamps, transforming into 73, 74
classification models
  about 41, 119
  decision trees 126, 127
  linear models 120-122
  naïve Bayes model 124
  predictions generating, for Kaggle/
    StumbleUpon evergreen
    classification dataset 133
  training 130
  training, on Kaggle/StumbleUpon
    evergreen classification
    dataset 131, 132
  types 120
  using 133
cluster 197
clustering evaluation
  URL 216
clustering models
  about 197
  hierarchical clustering 203
  K-means clustering 198-201
  K, selecting through
    cross-validation 217, 218
  mixture model 203
  parameters, tuning for 217
  training 208
  training, on MovieLens dataset 208, 209
  types 198
  use cases 198
  used, for making predictions 210
cluster predictions
  interpreting, on MovieLens dataset 211
collaborative filtering
  about 85, 86
  matrix factorization 86
comma-separated-value (CSV) 21
**components, data-driven machine learning
    system**
  about 42
  batch, versus real time 47
  data cleansing 43, 44

  data ingestion 42
  data storage 43
  data transformation 43, 44
  model deployment 45
  model feedback 46
  model integration 45
  model monitoring 45
  model training 45
  testing loop 45
content-based filtering 85
convergence 199
corpus 248
correct form of data
  using 147, 148
cross-validation
  about 45, 156-159
  K, selecting through 217, 218
  URL 157
customer lifetime value (CLTV) 161
customer segmentation 40

D

data
  exploring 55-57
  features, extracting from 70, 128, 164, 204
  movie dataset, exploring 62, 63
  processing 68
  projecting, PCA used 239, 240
  rating dataset, exploring 64-67
  transforming 68
  user dataset, exploring 57-61
  visualizing 55-57
data cleansing 43, 44
data-driven machine learning system
  components 42-48
data ingestion 42
datasets
  accessing 52, 53
  MovieLens 100k dataset 54, 55
data sources
  Amazon AWS public datasets 52
  Kaggle 53
  KDnuggets 53
  UCI Machine Learning Repository 52
data storage 43

data transformation 43, 44
decision trees
  about 126, 127, 154, 176
  impurity, tuning 154, 155
  tree depth, tuning 154, 155
  used, for regression 163
derived features
  about 73
  timestamps, transforming into categorical features 73, 74
dimensionality reduction
  about 221
  clustering as 224
  PCA 222
  relationship, to matrix factorization 224
  SVD 223, 224
  types 222
  use cases 222
dimensionality reduction model
  data projecting, PCA used 239, 240
  evaluating 242
  k, evaluating for SVD 242, 244
  PCA and SVD, relationship between 240, 241
  PCA running, on LFW dataset 235
  training 234
  using 238
distributed vector representations 274
divisive clustering 203
document similarity
  with 20 Newsgroups dataset 268-270
  with TF-IDF features 268-270
DStream
  about 282
  actions 284

E

EC2 Spark cluster
  launching 31-35
Eigenfaces
  about 236
  interpreting 238
  URL 236
  visualizing 236, 237

Elastic Cloud Compute (EC2) 8
ensemble methods 45
evaluation metrics 106
explicit matrix factorization 86-90
external evaluation metrics 216

F

face data
  exploring 226, 227
  visualizing 228
facial images, as vectors
  extracting 229
  feature vectors, extracting 232, 233
  grayscale, converting to 230, 231
  images, loading 229, 230
  images, resizing 230, 231
false positive rate (FPR) 138
feature extraction
  packages, used for 82
feature extraction techniques
  feature hashing 249, 250
  term weighting schemes 248, 249
  TF-IDF features, extracting from 20 Newsgroups dataset 251-253
feature hashing 249, 250
features
  about 70
  categorical features 70-72
  derived features 73
  extracting 92
  extracting, from bike sharing dataset 164-167
  extracting, from data 70, 128, 164, 204
  extracting, from Kaggle/StumbleUpon evergreen classification dataset 128-130
  extracting, from LFW dataset 225, 226
  extracting, from MovieLens 100k dataset 92-95
  extracting, from MovieLens dataset 204, 205
  normalizing features 80
  numerical features 70, 71
  text features 70, 75, 76

features, extracting
  feature vectors, creating for decision
      tree 169
  feature vectors, creating for linear
      model 168, 169
features, MovieLens dataset
  movie genre labels, extracting 205, 206
  normalization 207, 208
  recommendation model, training 207
feature standardization,
      model performance 141-143
feature vectors
  about 128
  creating, for decision tree 169
  creating, for linear model 168, 169
  extracting 232, 233

G

generalized linear models
  URL 121
general regularization
  URL 153
grayscale
  converting to 230-232

H

Hadoop Distributed File System (HDFS) 9
hash collisions 250
hierarchical clustering 203
hinge loss 123

I

images
  loading 229, 230
  resizing 230-232
implicit feedback data
  used, for training model 98
implicit matrix factorization 90
initialization methods,
      K-means clustering 202
internal evaluation metrics 216
inverse document frequency 248

item recommendations
  about 102
  similar movies, generating for MovieLens
      100k dataset 103, 104

J

Java
  Spark program, writing in 24-28
Java Development Kit (JDK) 9
Java Runtime Environment (JRE) 9
Java Virtual Machine (JVM) 8

K

k
  evaluating, for SVD on LFW
      dataset 242-244
K
  selecting, through cross-validation 217, 218
Kaggle
  about 53
  URL 53
Kaggle/StumbleUpon evergreen
      classification dataset
  classification models, training on 131, 132
  features, extracting from 128-130
  predictions, generating for 133
  URL 128
KDnuggets
  about 53
  URL 53
K-means clustering
  about 198-201
  initialization methods 202
  streaming 305
  variants 203

L

L1 regularization 189-191
L2 regularization
  about 188
  URL 153
label 128
Labeled Faces in the Wild (LFW) 225

lasso 162
latent feature models 88
Least Squares Regression 162
LFW dataset
  data projecting, PCA used 239, 240
  Eigenfaces, interpreting 238
  Eigenfaces, visualizing 236-238
  face data, exploring 226, 227
  face data, visualizing 228
  facial images, extracting as vectors 229
  features, extracting from 225, 226
  k, evaluating for SVD 242, 244
  normalization 233, 234
  PCA, running on 235
linear models
  about 120-122, 149, 150, 175, 176
  iterations 151
  linear support vector machines 123
  logistic regression 122
  regularization 152, 153
  step size parameter 151
linear support vector machines 123
logistic regression 121, 122
log-transformed targets
  training, impact 180-183

M

machine learning 51
machine learning models, types
  about 41
  supervised learning 41
  unsupervised learning 41
machine learning system
  architecture 48, 49
  business use cases 39, 40
MAE 174
MAP
  about 113
  calculating 114
map function 20
MAPK
  about 109-112
  URL 109

matrix factorization
  about 86, 224
  Alternating Least Squares (ALS) 91
  explicit matrix factorization 86-90
  implicit matrix factorization 90
Mean Absolute Error. *See* **MAE**
Mean Squared Error. *See* **MSE**
Mesos 8
mini-batches 281
missing data
  filling 69
mixture model 203
MLlib
  used, for normalizing features 81
model
  deployment 45
  feedback 46
  fitting 121
  integration 45
  monitoring 46
  training, implicit feedback data used 98
  training, on MovieLens 100k dataset 96-98
model inputs
  iterations 96
  lambda 96
  rank 96
model parameters
  decision trees 154
  linear models 149, 150
  naïve Bayes model 155, 156
  parameter settings, impact for
    decision tree 192
  parameter settings, impact for
    linear models 184, 185
  testing set, creating to evaluate
    parameters 183, 184
  training set, creating to evaluate
    parameters 183, 184
  tuning 148, 183
model performance
  additional features 144-146
  comparing, with Spark Streaming 306-310
  correct form of data, using 147, 148
  feature standardization 141-144
  improving 140, 177

model selection 45
model training 45
modern large-scale data environment
    requisites 37
movie clusters
    interpreting 212-215
movie dataset
    exploring 62, 63
movie genre labels
    extracting 205, 206
MovieLens 100k dataset
    about 54, 55
    features, extracting from 92-95
    movie recommendations,
        generating from 99, 100
    similar movies, generating for 103-105
    URL 54
MovieLens dataset
    about 53
    clustering model, training on 208, 209
    cluster predictions, interpreting on 211
    features, extracting from 204, 205
    performance metrics, computing on 217
movie recommendations
    generating, from MovieLens 100k
        dataset 99, 100
MovieStream 38, 39
MSE 107, 108, 113, 173, 174

N

naïve Bayes model 124, 155, 156
natural language processing (NLP) 248
nominal variables 71
nonword characters 256
normalization
    normalize a feature 80
    normalize a feature vector 80
normalization, LFW dataset 233, 234
normalization, MovieLens dataset 207, 208
normalizing features
    about 80
    MLlib, used for 81
numerical features 71

O

online learning 47, 279, 280
online learning, with Spark Streaming
    about 298
    K-means, streaming 305
    streaming regression model 298, 299
    streaming regression program 299
online machine learning
    URL 280
online model evaluation
    about 306
    model performance, comparing with Spark
        Streaming 306-308
optimization 121
options, data transformation 68
ordinal variables 71
Oryx
    URL 90
over-fitting and under-fitting
    URL 153

P

packages
    used, for feature extraction 82
parameters
    tuning 140, 177
    tuning, for clustering models 217
parameter settings impact, for decision tree
    about 192
    maximum bins 194
    tree depth 193
parameter settings impact, for linear models
    about 184, 185
    intercept, using 191
    iterations 185
    L1 regularization 189-191
    L2 regularization 188
    step size 186, 187
PCA
    about 222
    and SVD, relationship between 240, 241
    running, on LFW dataset 235
performance, classification models
    accuracy, calculating 134-136
    AUC 138-140

evaluating 134
precision 136, 137
prediction error 134-136
recall 136, 137
ROC curve 138-140
performance, clustering models
evaluating 216
external evaluation metrics 216
internal evaluation metrics 216
performance metrics, computing on MovieLens dataset 217
performance metrics
computing, on bike sharing dataset 175
computing, on MovieLens dataset 217
decision tree 176
linear model 175, 176
performance, recommendation models
built-in evaluation functions, using 113
evaluating 106
Mean average precision at K (MAPK) 109-112
Mean Squared Error (MSE) 107, 108
performance, regression models
evaluating 173
MAE 174
MSE 173, 174
performance metrics, computing on bike sharing dataset 175
Root Mean Squared Log Error 174, 175
R-squared coefficient 175
personalization 40
precision, classification models 136, 137
precision-recall (PR) curve 137
prediction error, classification models 134-136
predictions
generating, for Kaggle/StumbleUpon evergreen 133
generating, for Kaggle/StumbleUpon evergreen classification dataset 133
making, clustering model used 210
predictive modeling 41
Principal Components Analysis. *See* **PCA**
producer application 287-290
pylab 55
Python
Spark program, writing in 28, 29

R

rating dataset
exploring 64-67
RDD caching
URL 19
RDDs
about 14
caching 18, 19
creating 15
Spark operations 15-18
Readme.txt file
about 164
variables 164, 165
recall, classification models 136, 137
receiver operating characteristic (ROC) 134
recommendation engines
benefits 84
recommendation model
about 84
collaborative filtering 85, 86
content-based filtering 85
item recommendations 102
model, training on MovieLens 100k dataset 96-98
training 96, 207
types 84
user recommendations 99
using 99
recommendations
about 40
inspecting 101, 102
regression models
about 41
decision trees, for regression 163
Least Squares Regression 162
training 170, 171
training, on bike sharing dataset 171-173
types 162
using 170, 171
regularization forms
L1Updater 152
SimpleUpdater 152
SquaredL2Updater 152
REPL (Read-Eval-Print-Loop) 12
reshaping 229
Resilient Distributed Dataset. *See* **RDDs**

RMSE 108, 113, 173, 174
ROC curve, classification models 138, 140
R-squared coefficient 175

S

Scala
  Spark program, writing in 21-24
Scala Build Tool (sbt) 21
Singular Value Decomposition. *See* SVD
singular values 223
skip-gram model 274
Spark
  about 37
  installing 8-10
  running, on Amazon EC2 30, 31
  setting up 8-10
  URL 7
Spark clusters
  about 10
  URL 11
Spark documentation
  URL 121, 163, 283
Spark, modes
  Mesos 8
  standalone cluster mode 8
  standalone local mode 8
  YARN 8
Spark operations 15-18
Spark program
  in Java 24-28
  in Python 28, 29
  in Scala 21-24
Spark programming guide
  URL 28
Spark programming model
  about 11
  accumulators 19, 20
  broadcast variable 19, 20
  RDDs 14
  SparkConf 11, 12
  SparkContext 11, 12
  Spark shell 12, 14
Spark project documentation website
  URL 9
Spark project website
  URL 9

Spark Streaming
  about 47, 281, 282
  actions 284
  input sources 282
  model performance,
    comparing with 306-310
  transformations 282
  window operators 284
Spark Streaming application
  analytics, streaming 293-295
  basic streaming application,
    creating 290-293
  creating 286, 287
  producer application 287-290
  stateful streaming 296-298
standalone cluster mode 8
standalone local mode 8
stateful streaming 296-298
stemming
  about 264
  URL 264
stochastic gradient descent (SGD) 149, 280
stop words
  removing 258-260
streaming data producer
  creating 299-302
streaming regression model
  about 298, 299
  creating 302-305
  predictOn method 299
  trainOn method 299
streaming regression program
  about 299
  streaming data producer,
    creating 299-302
  streaming regression model,
    creating 302-305
stream processing
  about 281
  caching, with Spark Streaming 285, 286
  fault tolerance, with Spark
    Streaming 285, 286
  Spark Streaming 281, 282
supervised learning 41
Support Vector Machine (SVM) 121

SVD
  about 223, 224
  and PCA, relationship between 240, 241

T

targeted marketing 40
target variable
  training on log-transformed targets, impact 180-182
  transforming 177-179
term frequency 248
terms, on frequency
  excluding 261-263
term weighting schemes 248, 249
testing loop 45
testing set
  creating, to evaluate parameters 183, 184
text classifier
  training, on 20 Newsgroups dataset 271, 272
text data 247, 248
text features
  about 75, 76
  extraction 76-79
text processing impact
  evaluating 273
  raw features, comparing 273, 274
TF-IDF
  used, for training text classifier 271, 272
TF-IDF features
  document similarity, used with 268-270
  extracting, from 20 Newsgroups dataset 251-253
TF-IDF model
  document similarity, with 20 Newsgroups dataset 268-270
  document similarity, with TF-IDF features 268-270
  text classifier, training on 20 Newsgroups dataset 271-273
  training 264, 266
  using 268
TF-IDF weightings
  analyzing 266, 267

timestamps
  transforming, into categorical features 73, 74
tokenization
  applying 255
  improving 256, 257
training set
  creating, to evaluate parameters 183
transformations
  about 282
  general transformations 283
  state, tracking 283
true positive rate (TPR) 138

U

UCI Machine Learning Repository
  about 52
  URL 52
unsupervised learning 41
user dataset
  exploring 57-61
user recommendations
  about 99
  movie recommendations, generating 99, 100

V

variants, K-means clustering 203
vector space model 249

W

whitespace tokenization
  URL 255
windowing 284
window operators 284
within cluster sum of squared errors (WCSS) 198
Word2Vec models
  about 247, 274
  on 20 Newsgroups dataset 275-277
word stem 264

Y

YARN 8

Thank you for buying
Machine Learning with Spark

About Packt Publishing

Packt, pronounced 'packed', published its first book, *Mastering phpMyAdmin for Effective MySQL Management*, in April 2004, and subsequently continued to specialize in publishing highly focused books on specific technologies and solutions.

Our books and publications share the experiences of your fellow IT professionals in adapting and customizing today's systems, applications, and frameworks. Our solution-based books give you the knowledge and power to customize the software and technologies you're using to get the job done. Packt books are more specific and less general than the IT books you have seen in the past. Our unique business model allows us to bring you more focused information, giving you more of what you need to know, and less of what you don't.

Packt is a modern yet unique publishing company that focuses on producing quality, cutting-edge books for communities of developers, administrators, and newbies alike. For more information, please visit our website at www.packtpub.com.

About Packt Open Source

In 2010, Packt launched two new brands, Packt Open Source and Packt Enterprise, in order to continue its focus on specialization. This book is part of the Packt Open Source brand, home to books published on software built around open source licenses, and offering information to anybody from advanced developers to budding web designers. The Open Source brand also runs Packt's Open Source Royalty Scheme, by which Packt gives a royalty to each open source project about whose software a book is sold.

Writing for Packt

We welcome all inquiries from people who are interested in authoring. Book proposals should be sent to author@packtpub.com. If your book idea is still at an early stage and you would like to discuss it first before writing a formal book proposal, then please contact us; one of our commissioning editors will get in touch with you.

We're not just looking for published authors; if you have strong technical skills but no writing experience, our experienced editors can help you develop a writing career, or simply get some additional reward for your expertise.